BUGGING

ME?

WHAT'S BUGGING ME?

Identifying and Controlling Household Pests in Hawai'i

JoAnn M. Tenorio

Gordon M. Nishida

UNIVERSITY OF HAWAI'I PRESS

HONOLULU

© 1995 University of Hawai'i Press
All rights reserved
Printed in Singapore

95 96 97 98 99 00   5 4 3 2 1

Library of Congress Cataloging-in-Publication Data
Tenorio, JoAnn, M., 1943–
What's bugging me? : identifying and controlling household
pests in Hawai'i / JoAnn M. Tenorio & Gordon M. Nishida
p.   cm.
Includes bibliographical references and index.
ISBN 0–8248–1742–7 (alk. paper)
1. Household pests—Control—Hawaii.   2. Insect pests—Control—Hawaii.
3. Household pests—Hawaii—Identification.
I. Nishida, Gordon M., 1943– .   II. Title.
TX325.R46   1995
648'.7—dc20   95–10158
                                      CIP

University of Hawai'i Press books are printed on acid-free paper
and meet the guidelines for permanence and durability of the
Council on Library Resources

Designed by Paula Newcomb

CONTENTS

INTRODUCTION 1
How This Book Is Organized 1

DISCOURAGING AND CONTROLLING UNWANTED GUESTS 3
Prevention: Have a Little Less Aloha 3
Home Remedies 7
Chemical Control 10
Biological Controls 16
Miscellaneous Controls 17
Commercial Controls 18

OUR ARTHROPOD GUESTS 19
Structure of Arthropods 20
Life Cycles 22
Where to Get More Information 23

HOW TO IDENTIFY WHAT'S IN YOUR HOUSE 27
Key A. By Signs, Damage, Habitat, or Part of House Found In 27
Key B. By Kind of Insect or Arthopod, with Associated Signs 30

ANTS 36
What Ant's in My House? 37
Ants versus Carpenter Ants versus Termites 41
Ant Biology 41
Ants as Our Friends 42
Detection and Identification of Ants 42
Discouraging Ant Invasions 44
Control of Ants 45

BED BUGS 52
Impact of Bed Bugs 54
Control of Bed Bugs 54

BEETLES 55
Stored-Food Pests 55
Wood and Paper Pests 63
Fabric, Carpet, and Hide Pests 65
Nuisances 69

BOOKLICE 71

BUGS 73

CARPENTER BEES 74

CENTIPEDES 77

COCKROACHES 80
Impact of Cockroaches 81
Prevention of Cockroach Infestations 82
Detection and Monitoring of Cockroaches 83
Control of Cockroaches 83

CRUSTACEANS 89

EARWIGS 91

FLEAS 93
Impact of Fleas 95
Prevention and Monitoring of Fleas 96
Control of Fleas 98

FLIES AND MOSQUITOES 102
Filth Flies 102
Control of Filth Flies 104
Other Flies 107
Mosquitoes 109
Mosquito Species in Hawai'i 109
Detection and Monitoring of Mosquitoes 110
Control of Mosquitoes 111

GRASSHOPPERS, KATYDIDS, AND CRICKETS 114

MILLIPEDES 118

MITES 120
Rat and Bird Mites   120
Stored-Products Mites   122
House Dust Mites   123

MOTHS 126
Grain and Cereal Moths   126
Clothes Moths   128
Moths Occasionally Found around the House   131

SILVERFISH 134

SPIDERS 136
Widow, Violin, and Cane Spiders   137
Other Spiders   143
Control of Spiders   147

SPRINGTAILS 149

TERMITES 150
Termites versus Ants   151
A Mutually Beneficial Relationship: Symbiosis   151
Swarms   152
What Termites Do I Have?   152
Inviting Conditions for Termites   156
Detection of Termites   157
Prevention of Termite Invasions   159
Chemical Control of Termites   162
Other Methods of Detection and Control of Termites   164
Biological Control of Termites   165

TICKS 166
Detection and Monitoring of Ticks   168
Control of Ticks   168

WASPS AND HONEY BEES 170
Wasps   170
Honey Bees   175

SELECTED REFERENCES 176
ACKNOWLEDGMENTS 177
INDEX 178

INTRODUCTION

Hawai'i is a wonderful place for visitors. Two-legged and many-legged visitors alike take advantage of our wonderful climate and our warm hospitality. Though many of these guests are just passing through and won't disturb our households, some become unwelcome pests.

When we first discover "bugs" in our houses, many of us are too quick to grab the bug bomb and blast the unsuspecting creatures. Or, if we find the animal hiding in our food or belongings, we throw everything out—suspected culprit and all. We usually don't bother to ask ourselves if it's a good guy or a bad guy and, instead, use a scorched earth policy in dealing with often innocent visitors.

In your weaker moments, have you ever wondered what the critter was? What it does? Whether it will affect you? Or whether it will reappear? Have you wondered if you should be concerned? Do you know how to make your home less hospitable to these unwanted "bugs"? How you can control them without harm to family or pets?

This book gives you a closer look at some of the creatures that may invade your house and provides help for dealing with them. By no means do we provide complete coverage of the potential insects and their kin that might appear in or around your home, but we have tried to include the more common species you might encounter in Hawai'i.

How This Book Is Organized

The first section of the book presents an overview of prevention, control, and monitoring techniques for household pests in Hawai'i. Specific techniques are covered under the individual groups and species. The next section contains an introduction to the structure of insects and other arthropods, simple "keys" to household pests, and discussion of specific groups and species. The species contained in this book are listed by group in alphabetical order to make it easier for most non-entomologists to find the problem species.

Most biting and stinging creatures found in the house have been covered in *What Bit Me?*, the companion volume to this book. However, many that affect households in Hawai'i are covered again here, with additional information and with an emphasis on prevention and control. For example, fleas and ticks are included in this book because they may be distributed throughout the house, and prevention and control often affect large areas in the household; lice and scabies mites are not covered again in this book because they are largely restricted to the human host. If you cannot find a medically important species in this book and want to know more about it, check *What Bit Me?*

How to Capture Insects

Many pest species around the house are quite small and some magnification may be necessary to identify them correctly. A hand lens, magnifying glass, or a dissecting microscope may be necessary to see what they really look like.

If you do not know what insect is causing you problems, you may want to collect it for identification. Collect more than one specimen if possible.

Relatively hard-bodied creatures like beetles, cockroaches, and flies can be put in pill bottles or film canisters and the container covered or stoppered with cotton. Kill the pest by freezing it in the container for a few days.

Soft-bodied creatures like spiders, caterpillars, maggots, and grubs may be put in a vial or bottle with a tight-fitting lid; fill with rubbing alcohol until the specimen is totally covered with the liquid. This will preserve it until you can take it for identification.

Tiny organisms such as mites and booklice can be picked up with a small brush moistened with alcohol and stored in alcohol as described above. An option is to stick them onto transparent tape for transport to the identifying agency; try not to crush the specimens against the tape, because it makes them much more difficult to identify.

A sample of any damaged material is helpful for identification purposes. See "Where to Get More Information" (p. 23) for agencies that can help you with identification.

DISCOURAGING AND CONTROLLING UNWANTED GUESTS

Thanks to refrigeration, better packaging, and better sanitation in food-processing plants, pests are not as common as they used to be in our kitchens and pantries. However, dried food products available in Hawai'i usually spend a longer time in shipment and storage than comparable products in many mainland areas, thus increasing their chances of becoming infested. Besides the kitchen, other areas of the house also provide perfect opportunities for pests: carpets, clothes, books, storage areas, moist bathrooms and basements, and the very material that most of our houses are made of—wood. Some pests will attack nearly all types of materials, from feathers to lead, and our habits and lifestyles often provide many other sources of nourishment and shelter for them. Moreover, the warm, humid weather in Hawai'i is ideal for insects to breed and develop year-round; we cannot count on a frost or freeze during the winter to kill off a generation, so we must fight pest insects and their relatives all year.

The most popular concept in arthropod control nowadays is called "integrated pest management." Essentially, this means that you use all the weapons at your disposal to control pests, not just one method (for example, chemicals). Critical parts of this technique are prevention, early detection, and continuous monitoring. Using specific methods that target the pest alone and not methods that merely poison the environment are encouraged. This approach usually results in monetary savings and a healthier and safer environment. We emphasize this integrated approach throughout this book.

Prevention: Have a Little Less Aloha

One way to discourage unwanted guests is to make it harder for them to find a meal or a hiding place or to disrupt their travel plans. Your goal is to reduce or eliminate their chances of moving in with you. The first line of control is PREVENTION, and the simplest of controls is SANITATION. These alone are sometimes effective.

Use the Following Anti-pest Strategies before Trying Chemicals

✔ **Clean Up.** Clean up food scraps, wipe up drippings, and refrigerate leftovers. Keep open packages of food in airtight containers or plastic bags. Even dust is used as food by some stages of some pests, and most pests prefer items that are not heavily used or are tucked away in little-trafficked areas. Sweeping, vacuuming, and mopping are important weapons in the six-legged wars. Keeping food and crumbs cleaned up are also important if you are using baits; baits are only effective if the insects can find little else to eat and are forced to investigate the bait.

✔ **Preserve.** Many types of foods we eat today have their origins in methods to protect against pests. Though more often appreciated for flavor, some of the older preservative methods offer protection against pests. These methods include smoking, salting, pickling, and rapid air drying.

✔ **Lock Out.** After shopping, check your paper bags and boxes for hitchhikers trying to get into your home; cockroaches and silverfish are notorious for this. Check for holes or damaged wrapping in any food you are buying. At home, look for cast skins (skin left after an arthropod molts), cocoons, and tiny holes in packages of rice, pasta, grains, and cereals. Look for suspicious activity and dead carcasses in spices. Throw away packages with holes or obvious infestations. Place dry foodstuffs into airtight containers.

✔ **Dry Up.** Many of the problems we have with pests are often traceable to moisture. Many arthropods require a source of water to survive. Others prefer living in damper situations. Still others are attracted to molds and fungi that grow in high-humidity situations. Fix leaks and eliminate areas of puddling. Check drains and gutters and remove blockages. For rooms that are unusually damp, such as basements or rooms partly below ground level, do not install rugs that can pick up and hold moisture and act as factories for molds and fungi. Do not store items in paper boxes or keep paper products in these damp rooms. Ventilate the room; if necessary install a dehumidifier. If the source of your problem is outside the house, check the drainage on your grounds. Be careful with your choice of plants and ground cover for landscaping; dense ground covers and water-loving plants hold water in the soil.

✔ **Keep Cool.** Refrigerate as much of your grains as you can. Refrigeration helps by placing a food item in a less desirable area (most insects prefer warmth to cold), the cold slows down the growth and development of any pests or potential pests (for example, eggs) already in the material, and the cold (and drier conditions) also reduces the growth of possible attractants such as molds and fungi. If you must display or leave an item out in the open, the price is eternal vigilance—you must continually check the item, looking for signs of pest invasion or damage.

✔ **Be Vigilant.** Look for signs of pests or their damage. Periodically move furniture to sweep or vacuum underneath. Fine powdery dust or small, pelletlike droppings suggest borers in the wood. In kitchens and bathrooms, blackish brown streaks, droppings, and egg cases are cockroach signs. Periodically air out and check items in storage for fresh damage. Check the outside of the house for new damage, entry holes, dry rot, and other problems that may attract pests. Look for earthen tunnels leading from the ground into your structure.

✔ **Protect Pets.** Keep pets in the house free of parasites, especially fleas and ticks. If you walk your pet, have a parasite-detection program.

✔ **Reduce Clutter.** Piles of paper, boxes, books, refuse, and other accumulations of your home life provide perfect homes for immature and adult stages of insects. Good housekeeping can do much to discourage unwanted pests. If an item is badly damaged or infested and you can bear to part with it, throw it out—it is a likely source of more infestation.

✔ **Eliminate Hideaways.** Sanitation outdoors is also important to prevent invasion of pests into the house. Eliminate piles of logs and lumber on the ground, hollow tile blocks, leaf litter, compost, stacked building materials, abandoned tires, abandoned lawn furniture, and other such things. These all attract and shelter creatures that eventually may move into the house. Keep crawl spaces under your home clean and aerated; this is important for termite control and for discouraging black widow and other spiders.

✔ **Tighten Up.** To discourage entry into homes, put screens on windows and sliding doors and keep them in good repair. Keep cracks and crevices sealed, including those under doors and around window screens. Caulk around plumbing in the bathroom and kitchen. These are all perfect entry points for centipedes and cockroaches.

✔ **Be a Poor Host.** Cockroaches come into houses looking for food, water, and shelter. Deprive them of these as much as possible. Check regularly under sinks and around the house for dripping pipes and faucets. Look for pipes with condensation and fix leaking toilets. Some appliances serve as dark, warm motels for cockroaches—thus refrigerators, stoves, portable radios, and hot water heaters may be sources of infestation. If possible, regularly move all appliances away from the wall and vacuum underneath. Clean accessible places in refrigerators and stoves. If your refrigerator has an overflow pan, periodically remove the base panel, remove the pan, and drain and clean it. If your house is not elevated off the ground, pest entry can be reduced by keeping the perimeter around the foundation of the house clear of organic debris and plants. Plantings close to the house are aesthetically pleasing, but they increase the moisture level, inviting insects; they also screen the base of your house from effective inspection.

✔ **Reduce Breeding Sites.** Outside, eliminate standing water when possible, such as that in plants and plant containers. Find and eliminate possible breeding sites like old cans and bottles, old tires, hollows in rocks, tree holes, and so forth. If mosquitoes are a problem in the yard and house, consult your vector control agent for biological or chemical control measures that can be used against the larvae, which live in water. Remove old bird nests and paper wasp nests (carpet and related beetles can live off the feathers and insects in nests, so that these nests become sources of infestation). Vacuum the house and clean the floors regularly to reduce breeding places. Keep opened packages of food in sealed containers or in the refrigerator/freezer.

✔ **Manage Garbage.** To combat flies, empty your indoor garbage and trash every day. Wrap and seal the garbage and use outdoor garbage cans with lids. Wash out and sterilize cans, because larvae and eggs may remain on the sides and bottoms of cans. When flies enter the house, a fly swatter or flypaper is better than using sprays.

✔ **Be Tolerant and Be Patient.** Resist the knee-jerk reaction to reach for the bug spray when you first see an insect in the house. Ask yourself what it is, how it got there, what it is doing, will it hurt you or your house, and, finally, whether you can tolerate it or how you can safely eliminate it. Make the changes and apply the safe controls outlined in this book for pests. Be patient while the controls do their work.

Home Remedies

Fumigation

If the item is not too large and not too thick, small-scale fumigation might work. To do your own fumigation, place the item in a sealable, air-tight container or double bag it in thick plastic bags. Add a fumigant such as paradichlorobenzene (see discussion under Repellents and Fumigants on p. 14) and seal. Be very careful with these chemicals, because some people are sensitive to the vapors. In case of extensive infestations, large or bulky items, or repeated infestations, consult a licensed pest control operator.

Freezing

Freezing is a good way to eliminate pests. If you have the space and the time, freezing an item in a regular freezer might do the trick. The infested material should be sealed in a plastic bag and left in the freezer for a period of at least 12 days to kill all stages of the pest, including its eggs. In Hawai'i, be careful when removing sensitive or fragile materials from the freezer, because condensation will likely occur on the item. A non-frost-free freezer is recommended because a frost-free refrigerator works by cycling back and forth between cooling and warming to prevent the buildup of ice. Some insects, if given the chance, can adapt to freezing by taking advantage of the warming cycle, developing cold-hardiness, and thus survive.

Heat

Heat can also be used to eliminate pests. For example, in lightly infested material that you wish to save, such as flour or other grains, spread the material in a thin layer on a baking tin and bake at 130°F. for about 30 minutes to kill the adults, larvae, and eggs. The possible negative effects of this method include scorching the material or physically changing its characteristics, such as flavor. Heating dog kibble in this manner for about 15 minutes does not seem to change the attractiveness of the food to the pet, nor have insect parts in the food ever harmed any of our pets. Heat might also be tried in place of small-scale fumigation. If an item is infested, double bag it in dark-colored garbage bags and leave it out in the midday sun for a few hours.

Monitoring Devices: Traps

Many companies make simple sticky traps for catching insects. They are usually small, rectangular cardboard boxes with bands of sticky material on the inside bottom. These do not usually have poison but may have an attractant bait. Cockroaches go in and cannot get out; they get stuck and eventually die.

Traps will not eliminate cockroaches, but they are useful for several reasons. They can help reduce populations, they can pinpoint where the cockroaches are, and they can tell you whether you have eliminated your populations with other control measures or whether it's time to treat again. Unfortunately, sometimes these traps catch animals you don't want to catch. Don't be surprised if you catch something you didn't bargain for—a lizard or gecko stuck in a trap eventually dries up or starves to death.

Also very useful are yellow sticky traps. These are yellow cards (a color that attracts many insects) covered with a sticky glue that traps insects when they land on the card. This is an excellent tool for monitoring population levels.

Predators and Parasites

Many natural predators or parasites of insect pests are found frequently in the house. These include cane (banana) spiders, ensign wasps, web-making spiders, jumping spiders, centipedes, and geckos. Rather than shoo these creatures outside or, worse, kill them, consider encouraging them in the house—they can help to reduce your pest populations.

Whether you know it or not, there is a full-scale war going on right under your nose with predators and parasites battling it out with your pests. Predators and parasites are very effective in reducing populations of pests. Of course, sometimes there are negatives associated with predators and parasites. Geckos, for example, eat large numbers of insects in the home, but they leave evidence of their presence in the form of droppings. Some predators are a mixed blessing; ants can be bothersome about the house, but some do kill and eat termites and can be efficient predators of other insects. People are unnerved by the sight of cane spiders or centipedes crawling around on the walls and floors, even though they are both effective predators. Too often, friendly fellows are mistaken for enemies, and these allies are sprayed or swatted. Insects and related animals unfortunately do not wear distinguishing uniforms, so you must learn to recognize those that are friendly and tell them from the bad guys; otherwise you could be making your pest situation worse.

Geckos. A note here about some of the most helpful predators we can have around the house. Several species of geckos occur in Hawai'i. One of the most common in and around the house is the house gecko (Fig. 1), which seems to thrive around human dwellings. It is found at light fixtures, in houses on walls and ceilings, in cupboards, behind refrigerators—anywhere that it can catch and eat insects and other arthropods. Geckos are extremely beneficial, feeding on large numbers of cockroaches, mosquitoes, ants, termites, moths, spiders, and other obnoxious pests.

There is a downside to living with geckos, of course. Some people don't like their noises—various chirpings, clucks, squeakings, and other vocalizations. For others, the messy and abundant droppings of the geckos are a housekeeper's nightmare. If you absolutely can't stand to listen to gecko calls or don't want to clean up their droppings, you'll have to tighten up screens, doors, windows, and cracks and crevices by which they gain entry to the house. Geckos are attracted to lights where insects gather, so use bug lights (see p. 18 under Miscellaneous Controls) around doorways and around the outside of the house. If they come into the house and you don't want them there, shoo them out or pick them up gently and move them outdoors. Avoid grabbing geckos by

Figure 1. House geckos are extremely efficient at catching and eating insects. This gecko is not wearing red toenail polish; the red is clusters of tiny, parasitic mites. (Photo by Gordon Nishida)

the tail, because it may come off. This is a protective mechanism that allows them to escape predators; the tail will grow back.

> ### Tips for Discouraging Geckos
>
> ✔ Tighten and repair screens; caulk cracks and crevices where geckos can enter your house.
>
> ✔ Use bug lights outside the house to keep bugs from attracting geckos or move lights away from the house.
>
> ✔ Search out and destroy gecko eggs in light fixtures and wherever there is room to deposit eggs. If you must kill geckos, use sticky cockroach/rodent traps. Insecticides are not recommended.

Ants. Ants are more often thought of as nuisances than as helpful. However, their overall benefit to humans in an urban situation outweighs whatever harm they do. Ants enrich and aerate the soil and clean up the environment by feeding on dead and decaying organic materials. They also prey on many harmful creatures: cockroaches, filth fly maggots, flea larvae, and their great enemy—termites. Unfortunately, when ants begin making themselves a little too much at home, we want to evict them. It would be better for us if we could tolerate them in our houses, especially when we have termite and cockroach problems.

Chemical Control

For some pest problems, sanitation and home remedies are not enough. Sometimes you simply must turn to chemicals. But with chemicals, keeping insects at bay can sometimes be a trade-off between effectiveness and toxicity; and to control bugs safely, you must know something about their habits.

Because we are all concerned about the environment and because some people are sensitive to chemicals or wary of their use, we have included nonchemical ways of attacking pests. Chemicals suggested here are relatively benign to people and the environment. We give an overview below of some of the more common chemical controls that can be used in the household. Others are discussed under the insect groups.

Precautions for Insecticide Use

All insecticides are potential poisons. If you must use them or other pesticides, take the following precautions.

✔ Store pesticides securely and safely, away from children and pets. Don't store them with other household items like soap, hand lotion, kitchen cleaners, and the like.

✔ Always read and follow label instructions precisely.

✔ Limit your exposure to the chemicals as much as possible. Use waterproof gloves when mixing chemicals and mix only in well-ventilated areas; use masks when applying insecticidal powders.

✔ Be careful about general spraying, especially around food, dishes, pots and pans, cribs, toys, pet dishes, and similar items.

✔ Discard the empty pesticide container safely and do not reuse it. If all the pesticide is not used up, do not pour it down the drain; such chemicals can disrupt the normal operation of your septic tank or the treatment plant and poison the environment.

✔ Keep the Poison Center number handy for quick reference.

For those who have exhausted all other possibilities and do not want to deal directly with chemicals themselves, or if the infestation is widespread or untraceable, consult a licensed professional pest-control operator.

Dusts and Powders

Several dusts are particularly useful against household insects. These include the sorptive dusts such as silica aerogels and diatomaceous earth, as well as a crystalline material made from borax called boric acid. All of these dusts, although relatively safe for humans and pets, must be treated as insecticides and kept away from food, children, and pets. Wear a dust mask and gloves when applying these compounds.

Sorptive dusts may be abrasive and scratch off the waxy layer of the insect cuticle or may remove the wax by absorbing it. Either way, the insect dies of dehydration (water loss).

Boric Acid. This stomach poison has long been used as an anti-insect substance. Borax-based materials are widely used against ants, cockroaches, fleas, silverfish, and other household pests. Borates kill insects by killing the essential microbes that live in the insect gut. Boric acid is the most widely used borate material and is most notable in its effectiveness against cockroaches. It does not repel roaches, which usually avoid insecticides, and insects do not appear to develop a resistance to it. Insects may feed directly on boric acid, may pick it up on legs and antennae and swallow it in the process of cleaning themselves, or may simply carry it back to their nest, eventually killing nest mates.

One of the disadvantages of its use in Hawai'i, however, is the high humidity; this may cause the boric acid powder to clump, reducing its effectiveness. Some boric acid powders have an anticaking agent to prevent this problem, and some have an electrostatic charge that the manufacturer claims helps the powder to stick to the pest (for example, Roach Prufe). If one product doesn't work, try another one. Be patient; it may take a week or so for you to see results. If you succeed in reducing cockroach populations, then see them reappear after awhile, it may be time to vacuum up the old dust and reapply fresh powder. If you see any cockroach egg cases or small cockroaches, it is definitely time to treat again.

Boric acid powder also should not be used around children or around food. However, humans and pets must eat a large amount to be harmed, so it can be relatively safely used around the house. You can get boric acid in baits (see below) or you can apply it yourself. Dust it with a bulb duster or a plastic bottle into cracks and crevices, under appliances, behind cabinets, under shelf paper, and into areas not accessible to pets or children. It is best to use a dust mask while dusting.

Silica Aerogels. These materials absorb the waxy coating of the insect cuticle and cause the insects to dehydrate and die. Dri-Die is safe for applications in the home and works against many insects in confined areas, including cockroaches and drywood termites. Dri-Die can be dusted on cats and dogs for flea control. Drione is similar, but also contains a small percentage of pyrethrin insecticides and has an electrostatic charge on the silica-aerogel particles that causes it to stick better to the insect cuticle. However, it is slightly more toxic than Dri-Die because of the added insecticide. Drione is effective against many kinds of household insects.

Diatomaceous Earth. This material is mined from fossilized silica shell remains of diatoms. Like the silica aerogels, it absorbs the waxy layer of insect cuticle, causing the insect to dry out. In addition, it works as an abrasive to rupture the insect cuticle and cause the cell contents to leak out. It is not toxic to mammals and can be eaten in small quantities. Thus, it can be used to treat grains destined for human consumption (grains should be rinsed before eating). It can be used alone or in combination with pyrethrins (for example, Diacide) against many of the same insects to which silica aerogels are directed. Many garden shops in Hawai'i sell diatomaceous earth as a soil amendment.

For safe and effective powders for use on pets, pyrethrum and pyrethrins are probably the first choice. See the discussion of these compounds under Sprays (below) and refer to the discussion of Fleas (below).

Baits

Baits are poisons mixed with attractants or with food. Baits are now available at many grocery stores and drugstores and are especially good for controlling cockroaches and ants. They are usually quite effective and can kill within a few days. They are safe for use in the home, because it is difficult for children and pets to get at the poison within. However, pets and infants may try to chew the plastic shell if the bait tray is accessible; be sure to put it in areas where pets and children cannot get at it.

It is important that baits be used together with sanitation. The insects must not be sidetracked by other available food, but should be offered only the bait station.

For cockroaches, try baits containing boric acid or hydramethylnon (for example, Combat). Baits containing these chemicals are also effective for some ants, but it is important either to know what kind of ant you have or to experiment with baits. This is because certain species prefer sugar or carbohydrates, some prefer protein, and some fats or oils. An ant seeking sugar will not be attracted to a bait with only protein. You can make your own bait by mixing a teaspoon of boric acid powder with a cup of sugar water or with peanut butter, depending on the preference of your ants (see page 47). Spoon the bait into small jars with holes poked in the tops; be sure to place the jars out of reach of children.

As discussed under Dusts and Powders (above), boric acid is an old and proven insecticide. Hydramethylnon is relatively new and works by keeping the insect from digesting its food so that it starves to death. It is effective in very small amounts, so baits with hydramethylnon are safer than even boric acid because they contain relatively small amounts of poison.

Sprays

Insect sprays have their place, but they must be used with care. If you must spray, use a spray containing pyrethrins or pyrenoids, because these are less hazardous to the environment and relatively nontoxic to warm-blooded animals. It might be helpful here to explain the terms pyrethrum, pyrethrins, and pyrenoids, because these are among the most popular insecticides used in the household.

Pyrethrum comes from the dried, powdered heads of chrysanthemum flowers. Pyrethrin compounds are the active ingredients in pyrethrum. Pyrenoids are synthetic compounds that are similar to pyrethrins, but more toxic to insects. The synthetics also last longer in the environment. This is good if you want more lasting control, but bad if you want minimal poisons in your environment and less chance of harm to humans and pets. However, pyrenoids break down much faster in the environment than other more powerful insecticides (for example, carbamates and organophosphates).

The mode of action of pyrethrins against insects is attack on the nervous system; they cause instant paralysis. Many insects can detoxify pyrethrin and recover after the initial knockdown. Thus, the synergist piperonyl butoxide is usually combined with pyrethrins to prevent insects from recovering. There is some question whether the synergist might make the insecticides more toxic to humans and pets, so use synergized pyrethrins with caution.

Pyrethrum and pyrethrins are available as powders and are among the safest and most effective insecticidal powders for pets. Pyrethrins and pyrenoids are also common as aerosol sprays.

Sprays of any kind should always be used in conjunction with the longer-term controls described above or the insects will return. Learn what sprays are good for and what they are not so good for. For example, sprays are not very useful for ants or ant trails because your spray reaches only the scouts or worker ants and not the nest, where most of the ants live. Baits are a better bet.

Buy sprays appropriate to what you are trying to control. Buy flying insect spray products for flies and mosquitoes, and ant and cockroach sprays for these insects. Read labels carefully before buying and using.

Repellents and Fumigants

Repellents are compounds that repel insects rather than kill them. Fumigants kill by giving off gas that suffocates insects. Some insecticides have both repellent and fumigant actions.

Deet. Keeping mosquitoes away on campouts, cookouts, hikes, and other outdoor activities in Hawai'i often requires a repellent. Repellents don't kill mosquitoes or flies, but keep them from landing on your body. They do not work well or at all on wasps and bees. Most insect repellents are based on a chemical nicknamed deet (if you really must know, the chemical name is *N,N*-diethyl-meta-toluamide). In the 40 years since deet was synthesized by the U.S. Department of Agriculture, nothing else has been found to be as effective against as wide a variety of crawling and flying pests.

Excessive use of deet, however, can pose some risk, so use it sparingly. Use products with lesser concentrations of deet on children (40% deet has been recommended by *Consumer Reports* for adults, 20% for children [March 1993]). We found creams, sprays, and lotions ranging from 7% to 95% deet in a popular local drugstore. Off! Skintastic is sold for use on children; it has about 7% deet. Always follow the directions on the product label. Apply smaller amounts at first and reapply if insects are not repelled. In Hawaiian heat and humidity or if you perspire heavily, you may need to apply the repellent more often than the label indicates. Don't apply to children's hands, because they are too likely to put fingers in their mouths.

Citronella. Citronella is an essential oil extracted from the grass *Cymbopogon nardus*. Citronella candles may be used outside as a repellent for some protection in a small area against mosquitoes. Some people like the aroma; mosquitoes apparently don't.

Naphthalene, PDB, and Camphor. These three materials are most often used for protection of garments and fabrics against clothes moths and beetles. They must be used within sealed containers to build up a level poisonous enough to work and to protect humans and pets from their vapors. They come in ball, flake, or cake form and can be broken up and scattered around the garments in the closed container. They all make clothes smell and leave toxic odors.

Naphthalene and PDB (paradichlorobenzene) are effective as repellents against arthropod adults and young larvae. At the average Hawai'i temperature and humidity, PDB works as a fumigant; naphthalene repels but is unlikely to kill pests. Naphthalene is made from coal tar and is toxic if eaten and potentially damaging if vapors are inhaled. There is evidence that wearing clothes stored in naphthalene is harmful to children. PDB is much more toxic than naphthalene and is readily absorbed into the body by breathing; it is stored in body fat. It is best to avoid using this chemical. Naphthalene is sold in stores as moth balls; PDB is usually found as moth flakes or moth cakes.

Camphor is distilled from the camphor tree from Asia. It is repellent to moths and is also effective as a fumigant for adults and larvae, though it is less effective than naphthalene or PDB. Although probably less harmful than the other two, camphor is now recognized as also toxic to humans.

If you use any of these chemicals, air out or clean clothing before wearing or hanging in the closet to minimize exposure to the chemicals.

Other. Avon's Skin-So-Soft bath oil has supporters who swear that it keeps mosquitoes away. This product has no insecticidal or repellent ingredients, and Avon makes no claims about its repellent properties. *Consumer Reports* (1993) tested the bath oil against mosquitoes, along with known repellents. Skin-So-Soft was found to be "ineffective." Other studies have reported some repellency against mosquitoes, but the repellent effect is short-lived. Compared to deet, which provides protection for many hours (greater than 10 hours for 35% deet), the Avon product will probably provide limited protection for only a brief period of time. Avon's Skin-So-Soft Moisturizing Suncare does claim to repel mosquitoes, fleas, and deer ticks and contains oil of citronella as a repellent. It was not tested.

Biological Controls

Using biological controls is basically taking advantage of a pest's natural enemies—parasites, predators, competitors, disease organisms, and so forth. Biological controls range from protecting natural enemies already in the environment to importing natural enemies and trying to establish them permanently. For example, many wasps and moths have been introduced into Hawai'i by state agencies for control of many insect and plant pests.

The culture and sale of insects for use in biological control is now a commercial enterprise. Unfortunately, because of necessary plant quarantine restrictions in Hawai'i, many biological controls available for sale in mainland areas are not available here.

Reviews on the release of general predators like ladybugs and lacewings are mixed. They feed on everything, even the good guys; they may not survive in their new environment; and they may not stay where you want them. The insect predators and parasites already present in your home and garden—dragonflies, centipedes, geckos, toads, wasps, spiders—cost you nothing and are your most effective "insecticides."

Miscellaneous Controls

Many devices are commercially available that claim to chase away, repel, or kill insects. A few of the most common are discussed below.

Ultrasound Devices

These are tiny, battery-operated devices, usually sold as pet collars, claimed to keep away fleas on pets using high-pitched sounds. The pet collar does not work unless fleas are extremely close (within several millimeters) to the sound-producing device; fleas will just move to another part of the pet where they are not irritated. Other devices using ultrasound are sold as mosquito repellers and allow you to adjust the pitch of the sound. There is no evidence that these work at all with mosquitoes. We tried one device meant to be worn on the wrist and sold for use against mosquitoes—it did not even slow down Hawaiian *Aedes albopictus* mosquitoes from biting an attractive (to mosquitoes, that is!) human. In our experience, ultrasound devices do not work.

Bug Zappers

Bug zappers are black lights surrounded by a grid of electrified mesh. Insects are attracted to the ultraviolet lights and fly into the grid, electrocuting themselves. These devices would probably be more effective if it were pitch black all the time, because ultraviolet light sources do not compete well with the moon and mercury-vapor street lamps (almost all the street lamps nowadays use the lower wattage sodium-vapor lamps). Most important, the target insects, usually mosquitoes or other flies, do not always come all the way into the light. They come part way in, then sit at the fringe of the light. The net effect of the light is to attract insects to where you are—near your house, your pool, your barbecue, or other outdoor activities. Insects most attracted to the lights and zapped tend to be moths, and moths do not bite you.

Bug Suckers

These devices have a small motor connected to a tube and some sort of container at the end to trap insects. This allows you to "vacuum" the arthropod into a collecting bag and release it elsewhere or dispose of it. This may be a reasonable alternative for some insects and spiders, but bug suckers are probably not too effective on larger creatures such as centipedes, or those that can move very rapidly or fly quickly, or those that hide behind furniture and other places that can't be reached easily. If you're not interested in staring the "enemy" in the eyes, just use a regular canister vacuum.

Bug Lights

Though not 100% effective, a bug light used where lights are necessary on the outside of your house will reduce the number of insects attracted to your home. These lights send out a different spectrum, one that is shifted toward infrared, which insects are *not* attracted to, and shifted away from ultraviolet, which they *are* attracted to. On warm, humid, windless evenings when the whole world seems to be crawling with insects, you can minimize the attractiveness of your home by using bug lights outside and turning off all unnecessary inside lights. Even the light from the TV can be an attractant!

Commercial Controls

Fumigation is effective for drywood termites and many other household pests. The chemical almost universally used in Hawai'i for fumigation is Vikane (sulfuryl fluoride). Vikane is a tasteless, odorless gas that has tremendous penetrating power and is very effective in controlling pests. However, its effects do not last very long, and as soon as it is safe for humans to move back in, it is safe for pests to move back in, too. You must thus have a plan for prevention in place right after you fumigate. Several chemicals are used for ground treatment of termites. Chlordane was the treatment of choice previously, but it has been banned because of its potential health hazard. Chemicals now allowed for use are not as long-lasting nor as effective as chlordane, so most termite companies offer a much shorter guarantee period than previously. Homeowners should keep a watchful eye on their property and as soon as they detect a termite invasion, nip it in the bug! See the termite chapter (p. 150) for further discussion of termite control in Hawai'i.

OUR ARTHROPOD GUESTS

it wont be long now it wont be long
man is making deserts of the earth
it wont be long now
before man will have it used up
so that nothing but ants
and centipedes and scorpions
can find a living on it
— DON MARQUIS

There are far more arthropods on earth than all other animal species combined—over a million species named (and an estimated 10–30 million without names!). "Arthro" means jointed and "pod" means foot, so arthropods are joint-footed; that is, their legs are segmented into parts that are hinged together. Along with the presence of an exoskeleton (a skeletal structure on the outside of the body rather than on the inside), jointed legs are their major characteristic.

Five classes of arthropods live today: insects, arachnids (spiders, mites, ticks, and scorpions), millipedes, centipedes, and crustaceans (including sowbugs, pillbugs, shrimp, crayfish, lobsters, and crabs). Other arthropods, such as the trilobites, are long extinct. Among these classes, the insects are clearly the most dominant life form on earth—with nearly a million species described. With such an enormous number of species, the first task in recognizing individual species is to group them by their similarities and differences. Classification of these animals (in biological terms, their taxonomy) is thus based on arranging them into closely related groups.

The chart on p. 24 shows some of the higher categories for the arthropods included in this book and important features used to group them together.

The most important category, and the key to recognition and classification, is the species. A species includes all individuals that have similar form, physiology, and behavior and can be crossed to produce fertile offspring. A species always is identified by two Latin names, the first of which is the genus to which the species belongs. For example, *Latrodectus mactans*, the southern black widow spider, belongs to the genus *Latrodectus*. Notice that the genus is always capitalized and the species name is always lower case.

Other species closely related to the southern black widow spider also belong to this genus (for example, *Latrodectus geometricus,* the brown widow spider). Related species are grouped in genera, related genera in families, related families in orders, and so on. For example, the full classification for the southern black widow spider as compared with that of humans is as follows:

Common Name	Southern Black Widow	Human
Kingdom	Animalia	Animalia
Phylum	Arthropoda	Chordata
Class	Arachnida	Mammalia
Order	Araneae	Primata
Family	Theridiidae	Hominidae
Genus	*Latrodectus*	*Homo*
Species	*mactans*	*sapiens*

Structure of Arthropods

As explained above, arthropods are joint-legged. They also have exoskeletons (hardened coverings) that surround the body fluids and internal parts and provide support for the attachment of muscles and organs. Arthropod bodies are segmented and contain an open circulatory system, one without a system of veins and arteries. A simple brain and often a nerve bundle (ganglion) associated with each body segment usually make up the nervous system. Body types of the major arthropods are shown in Diagram 1.

Though covered with a hard outer shell, most insects are not sluggish and slow-moving or nonmobile like clams, snails, and barnacles, for example. Insects are both armored and very active, thanks to their "segmenta-

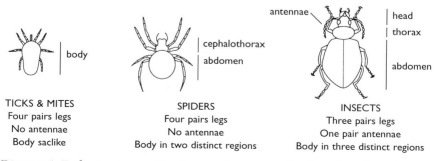

Diagram 1. Body structures of the three major groups of arthopods: ticks and mites, spiders, and insects.

tion." The body is divided into a series of ringlike pieces, all connected but each free to move on its own. Some worms (for example, earthworms) also have a segmented body, but spiders, insects, and other arthropods differ from worms in having jointed appendages on some body segments. Thus, worms are different from the young of insects that may look like worms (for example, grubs and caterpillars).

Arthropod Appendages Important to Identification

Antennae. Antennae are specialized, jointed appendages on the head. They are used for various purposes: feeling and sensing the surroundings, finding food, locating mates, detecting danger, and communication. Insects have one pair of antennae; arachnids (spiders, mites, ticks, and scorpions) have none; crustaceans (crayfish, crabs, and amphipods) have two pairs; and myriapods (centipedes and millipedes) have one pair like insects.

Legs. We have already said that the most characteristic thing about arthropods is their jointed legs in the adult stages; these are also usually present in immatures, though many, like maggots, are legless.

Insects have six legs in three pairs; arachnids have eight legs in four pairs (young often have six legs in three pairs); crustaceans have 10 legs in five pairs; centipedes have one pair per body segment; and millipedes have two pairs per body segment (sometimes over 200 pairs!).

Wings. Only insects among the invertebrates (groups without backbones) have wings. Among the insects, only full-grown adults can fly. Some adult insects are wingless. These include the silverfish and springtails, which are primitive groups whose ancestors probably never had wings. Other wingless groups like fleas are considered to have lost their wings in adapting to a parasitic lifestyle. Some kinds of termites and ants have wings that break off after they swarm. Most insects that fly have two pairs of wings, while others, notably flies, have only one pair.

Wings are important structures in insect classification. Most of the orders of insects end in "-ptera," meaning wing. Flies are *Di*ptera, or two-winged; butterflies and moths are *Lepid*optera, scale-winged; wasps and bees are *Hymen*optera, membrane-winged; earwigs are *Derm*aptera, skin-winged; grasshoppers and crickets are *Orth*optera, straight-winged; and so on.

Mouthparts. Mouthparts are one of the most important structures studied by entomologists (insect specialists). Mouthparts are adapted for the way an insect feeds and on what. To humans, the most important way insects cause damage and injury is by feeding. Knowing about an insect's mouthparts can help you tell what kind of damage it can do or has done; you can also look at damage to an object (for example, paper) and tell what kind of insect did the damage.

There are several types of mouthparts. The following are the most common, but there are many variations on these common types.

Chewing—These mouthparts work in tearing off, chewing, and swallowing, like a horse might eat grass. This is the basic type and is found in many groups: cockroaches, grasshoppers, caterpillars, beetles, silverfish, termites, booklice, and others.

Sucking/Piercing—Two operations are at work here. The first is to puncture or pierce the skin of an animal or the epidermis of a plant. The second is to suck the blood or liquid. These arthropods take in liquids like we suck juice through a straw. Sucking/piercing mouthparts are found in mosquitoes, mites, ticks, and many plant pests.

Sponging—The mouthparts are formed like a fleshy proboscis, which is retractable. House flies are the best example of this type. Flies feed on liquids such as nectar and sap, sponging them up. Flies can also dissolve some substances, like sugar, in their saliva and then lap it up.

Siphoning—Adult moths and butterflies have these very specialized mouthparts, which consist essentially of a long, slender, hollow tube held coiled up under the head when not in use. The moth straightens the coil to feed, puts the tip into exposed liquid, like nectar in a flower, and sucks it up.

Chewing/Lapping—This is an intermediate type found in the honey bee. Some parts serve chewing functions and some parts are formed into a lapping tongue.

Life Cycles

In controlling pests, we must understand an arthropod's life cycle. Often different techniques or tactics must be used on different stages because they live in different places and eat different foods.

Arthropods usually begin as eggs (though some are born alive) and hatch into an immature stage. As a result of their exoskeleton design, arthropods must molt (shed their skin) to grow. Several molts, or immature stages, may occur before the arthropod reaches adulthood. In insects, two different patterns occur in growth and development. In hemimetabolous insects, the young look like the adults, but don't have wings. These young, called nymphs, often compete with the adults for the same type of food. Insects in this category include cockroaches and true bugs (including bed bugs). In holometabolous insects, the young do not look like the adult and usually are in a niche (the role of an organism in its habitat) quite different from that of the adult. The young, called larvae, enter one last stage, the pupa, before becoming adults. Insects in this category include beetles, flies, bees, moths, butterflies, and wasps.

Where to Get More Information

The following organizations or agencies will be able to help you with questions about arthropods, their identification, and control. For medical questions, you should consult a physician.

About Insects

Bishop Museum, Department of Entomology, 1525 Bernice St., P.O. Box 19000, Honolulu, HI 96817, ph. 848-4194.

Hawai'i State Department of Agriculture, Plant Industry Division, 1428 S. King St., Honolulu, HI 96814, ph. 973-9530.

Hawai'i State Department of Health, Vector Control Branch (O'ahu), 2611 Kilihau St., Honolulu, HI 96819, ph. 831-6767. Vector Control branches on other islands: Hilo, Hawai'i, ph. 933-4386; Kona, Hawai'i, ph. 322-0033; Maui, ph. 877-2451; Moloka'i, ph. 567-6161; and Kaua'i, ph. 241-3306.

University of Hawai'i at Mānoa, Agricultural Diagnostic Service Center, 1910 East-West Road, Sherman 134, Honolulu, HI 96822, ph. 956-6706. On the island of Hawai'i: 875 Komohana St., Hilo, HI 96720, ph. 959-9155.

U.S. Navy at Pearl Harbor, Entomology, ph. 471-3948.

About Poisoning

Poison Center, O'ahu, 1319 Punahou St., Honolulu, HI 96826, ph. 941-4411. This is the only poison center in the Pacific Basin.

INSECTS AND OTHER ARTHROPODS

PHYLUM ARTHROPODA
- Body segmented
- Paired, jointed appendages
- External skeleton

SPIDERS, MITES, TICKS
Class Arachnida
- Eight legs in four pairs (young are often six-legged)
- One or two body regions; if two, the front is called the cephalothorax (head plus chest) and the back is the abdomen
- No antennae

Spiders
Subclass Araneae
- Two body regions
- Body constricted where it joins the cephalothorax
- Body not segmented

Mites and Ticks
Subclass Acari
- One body region
- Body round or oval, baglike, not segmented

CENTIPEDES
Class Chilopoda
- Many legs, one pair on each segment
- Long, flat body with many segments
- One pair of antennae

MILLIPEDES
Class Diplopoda
- Many legs, two pairs on each segment
- Wormlike, cylindrical body
- One pair of antennae

AMPHIPODS AND ISOPODS
Class Crustacea
- Two pairs of antennae
- Five or more pairs of walking legs
- Leglike appendages on abdomen

INSECTS
Class Insecta
- Six legs in three pairs
- Three body regions: head, thorax, abdomen
- Usually one or two pairs of wings
- One pair of antennae

Silverfish or Bristletails
Order Thysanura
- Wingless
- Carrot-shaped, gray, scaly
- Slender filaments at end of abdomen

Springtails
Order Collembola
- Wingless
- Tiny; jump when disturbed
- Abdomen with forked jumping spring

Crickets, Grasshoppers, Katydids
Order Orthoptera
- Large insects, chewing mouthparts
- Leathery front wings held rooflike or flat on back
- Hind legs enlarged for jumping
- Young resemble adults, but wings are small

Cockroaches
Order Dictyoptera
- Hindlegs similar to middle legs, adapted for running
- Antennae long, slender
- Chewing mouthparts
- Young resemble adults, but wings are small

Earwigs
Order Dermaptera
- First pair of wings small, squarish
- Long, narrow body ending with forceps
- Young resemble adults

Termites
Order Isoptera
- Winged (two pairs), found swarming *or* wingless, pale, delicate, in wood
- Wings similar in shape and size
- Antlike except for thick waist
- Young resemble adults

Booklice
Order Psocoptera
- Wingless; chewing mouthparts
- Long threadlike antennae
- Tiny, yellowish, with large head
- Young resemble adults

True Bugs
Order Heteroptera
- Usually two pairs of wings
- Front wings with first half leathery and second half clear and filmlike
- Piercing/sucking mouthparts (beak)
- Young resemble adults

Beetles
Order Coleoptera
- Two pairs of wings, first pair thick and leathery
- Wings usually meet in a straight line down the back
- Chewing mouthparts
- Young (grubs) do not resemble adults

Moths
Order Lepidoptera
- Fuzzy wings with scales or hairs
- Large, conspicuous insects *or* tiny, delicate, buff-colored
- Young (caterpillars) do not resemble adults

Flies and Mosquitoes
Order Diptera
- Usually one pair of wings
- Second pair of wings formed into slender, knobbed structures (halteres)
- Sucking/lapping/piercing mouthparts
- Young (maggots, larvae, wigglers) do not resemble adults

Bees, Wasps, Ants
Order Hymenoptera
- Two pairs of wings or wingless
- Forewings larger than hindwings and usually hooked together
- Young (grubs) do not resemble adults

Fleas
Order Siphonaptera
- Small, wingless
- Body greatly flattened side to side
- Usually many spines on the body
- Legs expanded for jumping
- Young tiny and maggotlike, do not resemble adults

HOW TO IDENTIFY WHAT'S IN YOUR HOUSE

Keys are one of the tools used by entomologists to sort through the many characters that separate the numerous species of insects and other arthropods. Most keys are extremely detailed and difficult to understand unless you are familiar with the scientific terms and know something about the group already.

Below are two simplified keys that will help you identify insects or other arthropods causing problems in your house. Descriptions lead you to names of groups discussed elsewhere in the book. If one description is not true for the insect or damage you have found, go on to the next description. Note also that these keys do not include many insects and arthropods of medical importance that are covered in *What Bit Me?*

—By signs, damage, habitat or part of house found in
    Go to Key A, below

—By kind of insect or arthropod, with associated signs
    Go to Key B, p. 30

Key A. By Signs, Damage, Habitat, or Part of House Found In

This key is meant to suggest possibilities for further investigation. If a description fits, go to the group and page numbers listed and read further to see if it matches your insect or sign. This key does not include many insects, spiders, and other groups that are found throughout the house or that are not associated with any particular clue. If your creature does not fit in any of these categories, flies freely about the house, or crawls or jumps around on the floor or walls or ceilings, see Key B (p. 30).

By Signs or Damage

—Insect or arthropod present, wandering or flying in house
    Go to p.29 (By Habitat or Part of House…)

—Webs or webbing present
- —Open webbing around house
  Spiders, p. 136
- —Matted webbing in foodstuffs, carpets, furniture
  Moths, p. 126

—Small, fibrous, or bristly cases, or almond-shaped or boat-shaped, brown or gray cases found around fabrics or clothing, or hanging on walls
Moths, p. 126

—Brown or blackish egg-shaped or purse-shaped capsules found in dark places like kitchen drawers, cabinets, shelves, in boxes, and so forth
Cockroaches, p. 80

—Small dustlike frass, pellets, black droppings
Termites, p. 150
Cockroaches, p. 80
Carpenter ants, p. 48

—Fine powdery dust, usually dropping from furniture
Beetles, p. 55, 63

—Vomit streaks or dried dark spots on counters, walls, in cupboards
Cockroaches, p. 80

—Holes, galleries, or chambers in furniture or books
Beetles, p. 55, 63
Termites, p. 150

—Holes, galleries, or chambers in wooden parts of house
Termites, p. 150
Carpenter bees, p. 74
Carpenter ants, p. 48

—Holes, abrasions, scrapings of fabrics, books, paper, carpets, hides, and other materials
Beetles, p. 55, 63, 65
Moths, p. 126, 128
Cockroaches, p. 80
Silverfish, p. 134
Booklice, p. 71

—Small whitish (young) or brownish adults feeding in rice, grain, cereal products
Beetles, p. 55, 58
Moths, p. 126

—Wings scattered about on floors and other flat surfaces
Termites, p. 150
Carpenter ants, p. 48

—Biting pets or people (pests of people and pets)
    Fleas, p. 93
    Ticks, p. 166
    Mites, p. 120
    Bed bugs, p. 52
    Ants, p. 36
    True bugs, p. 73
    Mosquitoes, p. 109
    Centipedes, p. 77
    Spiders, p. 136
—Stinging people
    Wasps and honey bees, p. 170
    Carpenter bees, p. 74
    Ants, p. 36

By Habitat or Part of House Found In

—In fabrics, carpets, hides, furs, and similar materials (fabric, carpet and hide pests)
    Beetles, p. 65
    Moths, p. 128
    Cockroaches, p. 80
    Ants, p. 36
—In wood, furniture, paper, books, cardboard, and similar materials (wood and paper pests)
    Termites, p. 150
    Beetles, p. 63
    Carpenter ants, p. 48
    Silverfish, p. 134
    Booklice, p. 71
    Cockroaches, p. 80
—In foodstuffs (stored-food pests)
    —In flour, grain, cereal, spices, dogfood, pasta, and other dried food
        Beetles, p. 55, 58
        Moths, p. 126
        Booklice, p. 71
        Cockroaches, p. 80
        Ants, p. 36
    —In meat, cheese, other protein-based foods
        Beetles, p. 55
        Flies (cheese skipper), p. 108

—In garbage, trash, sewage, excrement (feces), and fermenting or rotting food (pests of waste products)
  —Wormlike larvae or grubs
    Filth flies, p. 120
    Vinegar flies, p. 107
    Sap beetles, p. 62
  —Adultlike, but wingless
    Cockroaches, p. 80
—In damp areas, on mold, in bathrooms, on sinks, on refrigerators (nuisances)
    Moth flies, p. 108
    Booklice, p. 71
    Cockroaches, p. 80
    Millipedes, p. 118
    Amphipods, Sowbugs, or Pillbugs, p. 89, 90
    Earwigs, p. 91
    Crickets, p. 115
    Springtails, p. 149
—Climbing walls (predators, parasites, nuisances)
  —Usually on outside walls
    Wasps, p. 70
    Moths, p. 133
  —Usually on inside walls, crawling on floors, and similar places
    Wasps (ensign), p. 171
    Centipedes, p. 77
    Millipedes, p. 118
    Cockroaches, p. 80
    Termites, p. 150
    Flies, p. 102

Key B. By Kind of Insect or Arthropod, with Associated Signs

This key gives brief descriptions of insects and other arthropods you might find in the home. Insects are divided into two groups: those that obviously have wings and fly and those that do not appear to have wings, including immature (larvae and nymphs) forms. You will find some groups in both sections: for example, termites have both winged and wingless forms, so termites are listed in both sections. Within each section, groups are given in alphabetical order.

See also the chart on page 24 if you need more help to figure out what kind of insect you have.

—With wings and able to fly: adults
>Section 1
—Without wings or wings not apparent: adults or wormlike young stages
>Section 2

Section 1. Adults with Wings

Ants. Antlike insects of various sizes, winged or not winged, yellow to black, slender waist. Crawl rapidly around on surfaces, either alone or in trails with other ants. Some (carpenter ants) very large ($3/8$–$1/2$"). Can be found in any part of the house but often in the kitchen
>p. 36

Beetles. Hard-shelled, roundish or flattish insects, with wings hidden under hard covers (first pair of wings are hard and meet in a line down the middle). Fly slowly and heavily. Often found in kitchens or pantries infesting grains or meats or found in books or furniture
>p. 55

Bugs. All bugs have a proboscis, or beak, that they use to penetrate their food and suck up fluids (most are vegetarians). Bugs may have wings that are held flat over the back or wings that form a tent over the back. The flat-backed bugs usually have half the wing leathery and half filmy. Some can jump well; most fly clumsily
>p. 73

Carpenter Bees. Large (1" or more), black or golden, bumble bee–like insects. Nest in wood of buildings, poles, fences. Bees can be seen flying rapidly around eaves and entering large, round holes in wood. Make a loud buzzing sound
>p. 74

Cockroaches. Light to dark brown or blackish insects, shiny, flattened top to bottom, fast-moving, spiny-legged. Usually seen at night, often in kitchens when lights are turned on. Newly molted cockroaches are whitish
>p. 80

Crickets. Large, short-bodied, dark brown to black insects, active at night, make singing noises, hop when disturbed. Usually found outside, but may invade the house and disturb people with their chirping
>p. 115

Flies. Two-winged, generally grayish insects that zip around the house with

a buzzing sound, frequently landing on horizontal and vertical surfaces, on human and pet bodies, and on food
    p. 102

Mosquitoes. Slender-bodied, long-legged, long-winged flies that fly about the house, making a humming noise. Land on bodies of humans and animals to bite and suck blood
    p. 109

Moths. Large, furry-looking insects with scales on wings, usually landing on walls; often attracted to lights
    p. 126
  Or
Tiny light-colored fuzzy insects running over fabrics, flying about in closets and bedrooms (clothes moths), or flying about in kitchens or pantries (grain and cereal moths)
    p. 126–128
  Or
Tiny brownish moths flying in bathrooms, laundry rooms, or in places where almond- or boat-shaped gray silken capsules occur
    p. 133

Termites. Small (up to ½"), dark brown, winged insects, with broad, thick waist, found swarming on windless, humid evenings, emerging from cracks and crevices, or running around the house; attracted to lights, often flying around TVs, lamps, and other lights
    p. 150

Wasps. Usually long, slender-bodied insects, with a narrow waist and four wings, found flying about the house or nearby, often landing on walls. May be yellow and black, black, brown, or metallic. May sting or bite
    p. 170

Section 2. Adults or Young Stages without Wings or Wings Not Apparent

Crustaceans: Amphipods, Sowbugs, and Pillbugs. Grayish, armadillo-like crustaceans (sowbugs and pillbugs) or small, shrimplike creatures common in damp places outside. They sometimes enter homes and are usually found in the damper parts of the household
    p. 89

Ants. Antlike insects of various sizes, yellow to black, slender waist. Crawl rapidly around on surfaces, either alone or in trails with other ants. Some (carpenter ants) very large (3/8–1/2"). Can be found in any part of the house but often in the kitchen
    p. 36

Bed bugs. Broad, brown insects, flattened from top to bottom, wingless, no bigger than 1/5". Bites humans at night and sucks blood, usually in the bedroom. Unique odor
    p. 52

Beetles. Hard-shelled, roundish or flattish insects, with wings hidden under hard covers (first pair of wings are hard and meet in a line down the middle). Fly slowly and heavily. Often found in kitchens or pantries infesting grains or meats or found in books or furniture
    p. 55, 58, 63
  Or
Small, dark brown, hairy, caterpillar-like insects (larvae). Feed on rugs, carpets, stuffed or preserved objects of animal origin. Make large holes concentrated in a few areas. Small beetles may be present (carpet beetles)
    p. 55, 65
  Or
Small, whitish, slender or C-shaped larvae, do not live within silk. Found in wood, grains, seeds, peas, beans, meats, and dried fruit
    p. 55, 58, 63

Booklice. Tiny (1/25–1/12") insects, milky or yellowish, wingless, soft-bodied, with a large head. May feed on paper, grain, starch, or mold in damp places
    p. 71

Bugs. Medium-sized, with a beak used to penetrate their food sources. Occasionally attracted to lights or the insects that gather at lights
    p. 73

Centipedes. Long, slender, light to dark brown, somewhat flattened, many-legged arthropods, moving quickly in a wavy motion over floors and walls
    p. 77

Cockroaches. Light to dark brown or blackish insects, shiny, flattened top to bottom, fast-moving, spiny-legged. Usually seen at night, often in kitchens when lights are turned on. Newly molted cockroaches are whitish
    p. 80

Crickets. Large, short-bodied, dark brown to black insects, active at night, make singing noises, hop when disturbed. Usually found outside, but may invade the house and bother people with their chirping
    p. 115

Earwigs. Dark-colored, hard-shelled insects with conspicuous forceps (pincers) at the rear of the body. Crawling on floors, in boxes, or in damp areas
    p. 91

Fleas. Small (1/16"), brown, wingless insects, flattened from side to side. Jump when disturbed. Bite humans usually about the legs, feet, and waist, usually when dogs or cats are present in the household
    p. 93

Flies. Maggots (wormlike larvae) found in rotting food, garbage, trash, feces, sewage
    p. 102

Millipedes. Cylindrical, brownish, many-legged, short-legged arthropods, found moving daintily across floors, often in damp areas; occasionally climb walls
    p. 118

Mites. Tiny, microscopic arthropods that crawl on skin and bite humans, causing skin eruptions or hives with severe itching and sometimes other side effects, usually from handling infested straw, flour, dried fruits, or other groceries or from bites from rat or bird mites invading the home (bird, rat, flour and meal mites)
    p. 120

Moths. Small, whitish, fragile-looking worms (caterpillars). May be found in dry foodstuffs among webbing (food and grain pests), or may feed on clothing, fabrics, feathers, and other animal products (clothes moths, meal moths)
    p. 126

Silverfish. Small (3/8–1/2"), wingless, shiny, grayish, carrot-shaped insects with long, slender threads sticking out in front and back of body. Usually run around rapidly
    p. 134

Spiders. Wingless, eight-legged arthropods, with body in two sections. May live in webs or may run and hop around on flat surfaces. A few species may bite humans
    p. 136

Springtails. Tiny (less than 3/16"), wingless insects that jump into the air when disturbed. Usually found in damp areas, such as bathrooms, laundry rooms, basements
    p. 149

Termites. Small (less than 1/2"), whitish or light yellow-brown, soft-bodied, wingless insects with dark mouthparts, found eating and tunneling into wood
    p. 150

Ticks. Dark brown, eight-legged arthropods (young are six-legged), seed-like body, very round and fat when full of blood. Usually found sucking blood on dogs, climbing up walls, or hiding under mats, wood, or similar materials
    p. 166

ANTS
Order Hymenoptera
Family Formicidae

*Go to the ant, thou sluggard;
consider her ways, and be wise*
— PROVERBS 6:6

42 species in Hawai'i, 0 native

Over 40 species of ants now live in Hawai'i, all widely distributed outside the islands. Most are not household pests but are important to the environment, because they spend their time outside recycling nutrients and eating other insects, including termites and other ants. In fact, some ants threaten native species because of their insect-eating habits. Others may encourage insect pests on plants by protecting them from parasites and predators in exchange for nectar. Ants may invade homes during heavy rains or during dry spells looking for food, water, and shelter. A few species make themselves at home all year round and become pests as roaming individual ants or marching hoards on walls, countertops, kitchen floors, and bathrooms. Kill a few and they seem to come back like a bad dream.

Keep in mind that a few ants in your house are not necessarily harmful. They help keep the house free of other pest insects, including young silverfish, clothes moths, and cockroaches, and clean up old food and debris caught in cracks and crevices that you may not be able to reach. Putting up with a few ants here and there may be good practice. Of course, if you have stinging or biting ants, getting rid of them becomes more important.

Although it is helpful to know something about the biology of ants, you probably don't need to know exactly which species of ant you have. Except for the carpenter ant and a few other species, control measures that work on one species will usually work on another species. However, it helps to know if your ant likes sweets or oils or meat to be sure that control is effective. Try our recommended control measures below. If these don't help, you might have your ants identified so that you know how to get rid of them—or contact a professional exterminator.

What Ant's in My House?

The following is a small sampling of ants you may meet in houses in Hawai'i. These are also summarized in the table on page 40.

Figure 2. The pharaoh ant supposedly received its name because it was found in Egypt and presumed to be one of the plagues at the time of Pharaoh. (Photo courtesy of Van Waters & Rogers Inc.)

Figure 3. The pharaoh ant is difficult to get rid of because it nests in hard-to-reach places; it feeds on many different types of food. (Photo courtesy of Van Waters & Rogers Inc.)

✔ **Pharaoh ant** *(Monomorium pharaonis)* (Figs. 2, 3). Tiny (1/16") yellowish ant with lots of small nests often inside walls, behind kitchen ranges, and other difficult-to-reach areas; it also nests outside. It likes sweets, protein, fatty foods, warmer temperatures, and higher humidity. People locally complain about being stung, though its stinger is tiny. However, this ant does bite. The pharaoh ant is a pest in some hospitals; its small size allows it to crawl through tiny cracks and crevices and get into operating rooms and dressing stores. Ants have been shown to harbor pathogenic bacteria on their bodies but have not been proven to carry disease organisms from place to place.

A relative of the pharaoh ant, *Monomorium floricola*, also occurs fairly commonly in homes. It tends to nest in holes in walls and in trees and plants.

✔ **Tiny yellow house ant** *(Tapinoma melanocephalum)*. A tiny (less than 1/16" long) ant with strongly contrasting coloration—a blackish brown head with pale, brownish yellow body and antennae. This ant is very fast moving; it usually lives in the ground or in trees, but occasionally becomes a pest in homes. It feeds mostly on sweets, but also on grease and protein. At one time it was a common household pest in Honolulu, but reportedly disappeared by the late 1940s; it reappeared later and still causes occasional problems.

✔ **Glaber ant** *(Ochetellus glaber,* previously known as *Iridomyrmex glaber).* This very small, bald, black ant that bites fiercely is the most serious ant pest on Oʻahu today. It may nest in old termite holes in walls and roofs. It likes sweets but is also attracted to protein (meat).

✔ **Crazy ant** *(Paratrechina longicornis)* (Fig. 4). This small to medium-sized (up to ⅛" long) black ant can be recognized by its fast, erratic movements, changing directions suddenly and seeming to act "crazy." It is omnivorous and likes almost any food. A relative, *P. bourbonica,* is quite common, has similar biology, and is found on all the main islands.

Figure 4. The crazy ant moves in fits and starts, as if it can't make up its mind where to go. (Photo courtesy of Van Waters & Rogers Inc.)

✔ **Big-headed ant** *(Pheidole megacephala)* (Fig. 5). This is a medium-sized, dark brown ant so named because the soldiers have very large heads. It eats nearly anything; it is very aggressive and an efficient predator. At one time, it was the most common and serious pest in Hawaiʻi, and it has had a devastating effect on native invertebrates in the lowlands in the Hawaiian Islands. In its favored environment in the drier lowland areas, it is usually the dominant ant. It is usually only a nuisance in households.

Figure 5. The western bigheaded ant is similar to the bigheaded ant in Hawaiʻi. Soldiers have giant heads and jaws, but they are mostly stay-at-homes, protecting the nest. The tiny workers are the ones that go after and kill other insects such as termites. (Photo courtesy of Van Waters & Rogers Inc.)

✔ **Argentine ant** *(Linepithema humile,* previously known as *Iridomyrmex humilis)* (Figs. 6, 7). This medium-sized (⅛" long), reddish black to tan ant used to be the worst pest in Hawaiian homes, but may be being replaced by the glaber and pharaoh ants. It usually nests outdoors, but often invades homes looking for food. It is reportedly a vicious "stinger," but the ant, in fact, does not have a stinger. Instead, it bites and sprays the wound with a chemical that produces a sharp, burning sensation. It prefers sweets.

Figure 6. Argentine ant workers carrying pupae to safety. This ant has very poor eyesight and literally bumps into its food. Once the food is found, a chemical trail is left for other ants to follow back to the food source. (Photo courtesy of Van Waters & Rogers Inc.)

Figure 7. The Argentine ant has a sweet tooth, and sugar is a favorite food. (Photo courtesy of Van Waters & Rogers Inc.)

Figure 8. The Mexican ant is usually found outdoors and will often sting to defend the plants or insects it is tending. (Photo by Neil Reimer)

✔ **Mexican ant** *(Pseudomyrmex gracilis mexicanus)* (Fig. 8). A medium-sized, brown and black, wasp-like ant found on Oʻahu, especially in drier areas, notably Hawaiʻi Kai and around Koko Crater. The stings are painful, causing burning or throbbing around the sting site. It nests in hollow branches outside. This ant is attracted to meat.

✔ **Longlegged ant** *(Anoplolepis longipes)*. This medium large, yellow to reddish yellow ant is very fast moving and is usually found outdoors. It likes sweets and meat. At certain times of the year, this ant may invade homes in large numbers, but it is only a short-term nuisance; it does not sting or bite.

✔ **Carpenter ant**. See the discussion below and on page 48.

Some Ants Commonly Found in Houses in Hawai'i

NAME	FEATURES	OILS/FATS	SWEETS/CARBOHY	MEAT/PROTEIN	INSIDE	OUTSIDE	IMPACT	ISLAND[a]
Pharaoh ant *Monomorium pharaonis*	Tiny, yellow		✓	✓	✓	✓	Bites	O,M,H
Monomorium floricola	Tiny, black head & abdomen, lighter middle		✓	✓	✓	✓	Bites	All main
Tiny yellow house ant *Tapinoma melanocephalum*	Tiny, fast moving, black head, pale body		✓		✓	✓	Nuisance	All main
Glaber ant *Ochetellus glaber*	Small, black		✓	✓	✓	✓	Bites	O,M
Crazy ant *Paratrechina longicornis*	Small-medium, erratic moving	✓	✓	✓		✓	Nuisance	All main
Big-headed ant *Pheidole megacephala*	Medium, dark brown, soldiers with large heads	✓	✓	✓		✓	Nuisance	All main
Argentine ant *Linepithema humile*	Medium, reddish black		✓			✓	Bites/Sprays	All main
Mexican ant *Pseudomyrmex gracilis mexicanus*	Medium, wasplike, brown/black			✓		✓	Stings	O
Longlegged ant *Anoplolepis longipes*	Medium-large, reddish yellow, fast moving		✓	✓		✓	Nuisance	K,O,M,H
Carpenter ant *Camponotus variegatus*	Large		✓	✓	✓		Bites	All main

Source: Adapted from Neil J. Reimer, *Termite Times*, Dec. 1993, no. 4.
[a] O, O'ahu; M, Maui; H, Hawai'i; K, Kaua'i. "All main" includes Moloka'i and Lāna'i.

Ants versus Carpenter Ants versus Termites

Carpenter ants are quite large (5/16–3/4") and reddish brown. Carpenter ants in Hawai'i chew into wood to make nests but do not feed on the wood. They also nest in clocks, pianos, cardboard boxes, suitcases, and many other such hideaways. They are active at night, unlike most other ants, which are active during the day. Winged forms also swarm in the process of founding new colonies. This leads to confusion with termites, which also come out at dusk to swarm. Piles of broken-off wings may mean you have carpenter ants or termites (see Diagram 2). Though much larger than other ants, the carpenter ant is similar to other ants in having a narrow waist, elbowed antennae, and two pairs of wings of unequal size. It also often has dark brown stripes across the top of the abdomen.

See the discussion on page 151 to be sure you aren't confusing termites and ants.

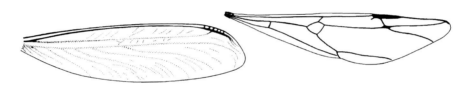

Diagram 2. Termite wings *(left)* are rounded and have many veins, while ant wings *(right)* are angular and have fewer veins.

Ant Biology

Like honey bees and termites, ants are social insects with specialists that have specific duties within the colony.

Queens. There are one or more in a colony: some species have one, and some have several queens. The queen uses her wings during the mating flight only, then sheds them and seeks a nesting place. In some species, the queen starts a new colony. In other species, she may return to the old colony to lay eggs and there will be multiple queens in the colony.

Winged Females. These have the potential of becoming queens.

Winged Males. These only exist to mate with and fertilize females. After the mating flight, they die—their job is over. The queens, winged females, and winged males are called reproductives.

Soldiers. These are found in some species but not all. Similar to workers, they defend the nest. Soldiers tend to be larger than workers and have bigger mandibles (jaws). Soldiers of some species (for example, *Pheidole megacephala*) have very enlarged heads.

Workers. Workers are the ant colony's lifeblood. These sterile females forage for food, feed the queen and larvae, and do most of the work to build, enlarge, and maintain the nest.

The queen lays eggs after establishment of a nest and mating. Eggs develop into larvae in about a month, and a pupal stage takes another 3–4 weeks before adults emerge. Queens live 12–15 years, workers only 3–4 months. The Argentine ant has many queens that lay hundreds of eggs. Larvae are tiny, blind, white, and shaped like crookneck squash; they are carried about by the workers, who feed them with regurgitated food. Sometimes, when the ground is flooded or if you break open the nest, you may see workers carrying the whitish larvae and pupae to safety.

Ants as Our Friends

Ants are bitter enemies of termites. Occasionally, depending on the species involved, you may see ants in your home because they are attacking termites that are there. In most cases the ants will have the upper hand over the termites. However, you probably should not depend on ants to eliminate your termite problem.

Ants also fight other ant species, one displacing the other depending on the circumstances. This is why you normally don't see many different species of ants at one time in one place; one species usually eliminates the others. Occasionally, ant species overlap at the same location because they have different habits and food sources.

Detection and Identification of Ants

You probably know by now that you are not dealing with a termite or a carpenter ant. At this point, you can try to identify which ant species you have or you can simply try to get rid of the ants by observing their habits. Don't think that simply wiping them off the counter will rid you of the pests; they will come right back unless you eliminate their source of food or seal off their access routes. Answers to the following questions will help in controlling ants:

1. What are the ants feeding on? Ants like three kinds of food: protein (meat), usually other insects; carbohydrates or sweets; and oils or fats. Ants eat a wide variety of foods but usually prefer one kind over another. You should know what attracts your ants. If there's a trail, set down different things alongside it to see what they like. Try sugar water, peanut butter as a protein (and fat!), and some kind of fat or oil. If they swarm over the bait, they like it; if they don't seem interested, they don't like it. Once you know what kind of food they like, you have accomplished two things—you know what foods to be more careful about leaving around and you know what the ants can be attracted to for use in a killing bait. For a carbohydrate-hungry ant, for example, powdered sugar and water with boric acid mixed in is a good bait. Keep in mind that even though an ant may prefer sweets, if the colony has already overindulged on sweets, a protein (or fat/oil) may work better as a bait. So you may need to experiment to find the right bait.

2. How are the ants getting into my house? You often see ants in the home moving in line, along a trail. This is because the first scouts to find food leave a chemical (called a pheromone) trail so that other workers can find the food source. If you watch ants carefully, you might see them touching the end of their abdomen to the ground as they walk, marking the trail; this is one of the ways that ants communicate. Workers follow this chemical trail, resulting in an almost militarylike formation of tiny ants marching along in single file. You can use this line of ants to backtrack and discover where ants are entering a room or to locate a nest to apply insecticide (see Control of Ants below). If you erase the trail by cleaning with detergent without killing the ants, they will wander about aimlessly until they pick up another trail.

3. Where is their nest? Nests may be inside or outside the house. Look first inside the house by following the ants. Lift up and examine anything the ants appear to be coming from. They usually prefer items of organic origin: knife blocks, wooden boxes, or under or in soil of house plants, for example. If they disappear into a wall or floor, they may be nesting inside the wall or coming from outside the house. Ants invading a house from outside can often be found nesting in surrounding areas, in or around trees and shrubbery. Survey the outside of the house, looking for trails of ants. Follow the trails, looking for signs of ant hills or disturbed soil or holes with lots of ants around. Ants may also be in potted plants or planter boxes. If you find the nest, you will see adult ants along with whitish larvae and pupae. Treat the nest as described below. If you can't find nests outside, the ants may be nesting inside the house.

Discouraging Ant Invasions

Storage of Food

Eliminating food for ants is the best way to prevent their establishment. Keep the kitchen in particular as clean and free of drippings and crumbs as possible. Food standing out in the open is an open invitation; store all food in sealed containers. Rinse out food packaging before throwing it into the wastebasket. If you have pet dishes in the kitchen, keep them clean and free of leftovers. If you follow these few rules, the population of ants can be kept to a minimum. And, every step you take to reduce the ant population reduces the possibility that you will have cockroaches, pantry pests, and other common household insects. Besides, reducing those pest ants reduces the number of their predators—centipedes and spiders, for example, that you also may dislike having in your home.

Waste Management

Ants will scavenge for food wastes in trash and garbage cans and wherever else the remains of human and pet meals accumulate. You must discourage ants from finding food. Rinse glasses, jars, paper and plastic containers, meat wrappings, frozen dinner containers, and other items before tossing them in the trash, especially if you will not be taking out the trash for awhile. Run your garbage disposal regularly or, if you don't have one, separate organic waste, such as food scraps, leftovers, spoiled fruits and veggies, and bones from other waste and keep it in sealed containers until disposed of.

Caulking and Tightening Up

If you follow an ant trail back to its source, you can often find where ants entered the house. Use a good grade of silicone caulk to fill any holes or cracks to bar access. You may have to do this several times because ants cleverly find alternate entries. If ants are coming in through a double-wall structure, they may be nesting between the walls. You can apply a boric acid dust or a desiccant such as silica aerogel (see below) through the holes before sealing.

Control of Ants

Physical Control

Sponging Up in Emergencies. What to do when ants are suddenly swarming over your pet food dishes or in your sink full of dirty dishes? You can opt for a temporary fix, followed by the long-term solutions discussed below. Keep in mind that killing individual workers won't have much effect on the population, but keeping food from them and sealing their entryways will over time eventually make inroads into the ant colony. Here's what to do to quickly regain control of your kitchen counter:

✔ Watch the ants to find out what they're after. Follow the trail back to see where they're coming from.

✔ Once you know where the ants are coming from, you can kill the ants you see. Leave the food in place so that the ants remain in line; they're easier to kill when they're orderly.

✔ Use a soapy sponge to wipe up the ants along the trail, rinsing occasionally in water. Or you can use a soapy spray or liquid detergent to squirt along the trail and wipe up the ants with a sponge. Any commercial kitchen cleaner will also work.

✔ Once the ants have temporarily stopped marching, block their point of entry with tape or petroleum jelly and remove the food source. Later, caulk the hole or crack to bar further entry.

Barriers. You can use barriers made from sticky materials or moats of soapy water to protect susceptible areas. If you're having a picnic outdoors on a picnic table, for example, and the ants have invited themselves to your picnic, set the legs of the table in containers of soapy water to prevent more ants from helping themselves to your lunch.

Chemical Control

Our first impulse is to grab an insecticide and spray the dickens out of ants, grinning with glee as we watch them die. The down side is that you get only a small percentage of the ant colony with sprays; typically only 10–20% of the colony is out foraging and never the queen. Have you ever considered how much you are paying for each tiny dead body? In some species, spraying may make the colonies fractionate, resulting in lots of little colonies and more trouble. Spraying broad areas inside or outside the house is not recommended for general ant control. It is expensive, bad for the environment, and works only for a short time. Because all natural enemies are killed off as well, chemical control is a trap that commits you to indefinite repeated sprayings. It is much better to use the dusts or baits described below; they are cheaper, last longer, and are better for the environment.

Dusts. Dusts such as silica aerogels and diatomaceous earth are discussed on page 12. Silica aerogels are chemically inert materials used as dehydrating (drying) compounds. You are probably familiar with the small packets of silica aerogels that accompany new cameras and electronic equipment. Silica aerogels made for insect control break down the waxy protective coating of an insect's cuticle, causing the insect to dry out.

Diatomaceous earth works similar to silica aerogels, but also scratches and punctures the insect cuticle (skin). It is nontoxic and can be used alone or in combination with insecticides like pyrethrins. These dusts can be blown into cracks and between walls. They retain their killing power for years.

Boric acid dust can also be used. It is a stomach poison and kills ants slowly over time after they eat it in baits or by grooming themselves after walking over the dust. It is widely available in stores, often in handy plastic applicators for convenient use.

For faster action, pyrethrins can be combined with dusts for quick knockdown. There are some good commercial preparations, both aerosol and nonaerosol, with applicators that allow you to apply the dusts finely to cracks and crevices without blowing the insecticide into the air and exposing yourself to toxic substances.

Be sure to follow label directions and take care not to breathe the dusts. Either have a professional apply them or for small-scale applications wear a dust mask and goggles.

Baits. We like the commercial bait "traps" for controlling ants because they are clean and nonmessy, safe, and convenient—compatible with a busy lifestyle and concern for the environment. An advantage of the baits is that you don't need to know where the nest is to use them. Baits work on the principle of attracting ants with food they like and incorporating in the food a slow-acting poison. Workers eat the bait and take it back to the colony to share. There are a number of poisons used in the bait stations: hydramethylnon, chlorpyrifos, arsenic, and boric acid. The poison used must not kill too quickly or workers will die before they can spread the poison through the nest.

There are two problems with baits for ants: (1) attractiveness—the ants must be attracted to the bait or the method will not work, and (2) the nature of an ant colony. Queens are much bigger than other ants and thus need a larger dose of poison and may not die for some time. Pupae and eggs will survive, because they have not eaten the bait, and workers will emerge as long as the queen lays eggs. The Argentine ant has many interconnected colonies so it's hard to wipe out completely. Thus, it is possible to reduce the size of a colony, but it is very difficult to eliminate it completely. You may have to reapply baits periodically. Do not use any other chemical spray or dust while using bait stations. Give the bait a chance to work over several days. If one brand doesn't work, try another with a different active ingredient or attractant. Other alternatives are "sweetening" the bait yourself or making your own bait.

Homemade Boric-Acid Bait

Use only boric acid formulated for pest control. It is dangerous to use medicinal boric acid because it can be confused with sugar or salt.

- ✔ Mix 1 1/2 teaspoons of boric acid powder with 1 cup of water plus 1/2 cup powdered sugar.
- ✔ Fill small jars with cotton balls and soak the cotton with the liquid.
- ✔ Poke several holes in the tops of the jars so ants can get in and out easily.
- ✔ Place the jars out of reach of pets and children.

The trick is to use enough boric acid to eliminate ants but not so much that the scout ants are killed before they can return to the colony. If many ants die around the bait station, there is too much poison. If ants keep coming in unreduced numbers for more than a week, your formula is probably too weak.

"Sweetening" a Commercial Bait. Once you know what your ants like (sugar? leftover cooking oil? peanut butter?), you can add a small amount of it by poking it inside the holes in the bait trap. Try a few drops of sugar water, oil, or a small dab of peanut butter. If you see ants entering and leaving the trap in larger numbers, you have succeeded at enhancing your bait trap.

Treating Nests Outside. If you find a nest outdoors, try drenching the nest area with soapy water or an insecticidal soap formulated to kill insects. Give it a couple of tries before deciding it didn't work. You can also try dousing the nest with boiling water or dusting it with diatomaceous earth. If these don't work, try a pyrethrin insecticide. You may not be able to kill all members of the colony, but you may succeed in getting the ants to move their nest farther away where they won't bother you. After all, the point is to get the ants out of your hair, not exterminate them.

Hawaiian Carpenter Ant
Camponotus variegatus
(Fig. 9)

The Hawaiian carpenter ant is found on all the major islands. It is the largest ant in Hawai'i, measuring from 3/8–1/2" long; it is usually yellowish brown, with dark horizontal stripes across the top of the abdomen. These ants don't sting but can bite painfully.

Yates (1988) covered the Hawaiian species well in the *Urban Pest Press* series (vol. 1, no. 2). You can find additional details on biology and control in that publication, available from the College of Tropical Agriculture and Human Resources at the University of Hawai'i.

Like other ants, carpenter ants are social, living in colonies. Sexual forms swarm during the summer in Hawai'i. After sealing herself in a small cavity, the queen lays 15–20 eggs. She feeds the new young on stored body fat until the young can care for the next brood. Then she becomes an egg-laying machine, leaving care of the colony to other ants. The period from egg to new adult worker is about 2½ months. As the colony grows, workers are

Figure 9. Carpenter ants are large and can bite. The ant on the right has wings and is the reproductive form that leaves the nest, mates, and establishes new colonies. The ant on the left is a worker. (Photo courtesy of Van Waters & Rogers Inc.)

produced in different sizes. Some take on the work in the colony and care for the immatures and reproductives; others defend the colony and search for food. After 3–6 years, the colony contains the one wingless egg-laying queen and up to 3,000 workers. At that time winged reproductives are produced to establish new colonies, and the cycle of swarming and colonies formation is repeated.

The carpenter ant is nocturnal and forages at night for food. Because it swarms in Hawai'i at about the same time of year as termites, it is sometimes confused with termites. You may see winged forms in your home after a swarm or you may see wingless workers at night looking for food. Refer to the figure on page 151 to distinguish between termites and ants; the size of this ant alone, coupled with its narrow waist, makes it easy to separate from termites and other ants. If you only find the wings, take a closer look at them (see Diagram 2, p. 41). If the wings have straight main veins with very few cross-veins, or with tiny cells at the wing tip, they are termite wings. If the wings have cross-veins, creating large cells, they are ant wings.

Despite its reputation as a wood pest and unlike its mainland relatives, the Hawaiian species does not feed on wood. It feeds on insects, sweet secretions (honeydew) from insects, as well as most foods in your home—fruit juices, meat, sweets, grease, and fat. Hawaiian carpenter ants may enlarge holes and tunnels they find, but they do little or no damage to sound wood structures in Hawai'i.

Detection of Carpenter Ant Nests

The Hawaiian carpenter ant makes nests in wood hollowed out by other insects, in dead parts of trees, and in any natural or artificial space it finds suitable. Nests inside the home can be found in wall clocks, pianos, between double walls, in little-used items in storage such as cardboard boxes, suitcases, drawers, debris, and similar places. Hollow-core doors found in many structures in Hawai'i are just to their liking.

Yates states that "six or more foraging workers indoors is a good indication" that carpenter ants have established a nest in your home. If you find that many, you should look for the nest. Simply spraying individual ants will do little. Try the following to find a nest inside the house:

✔ Start out in the room where you see most ants. Look around for any stash of trash or stored items that hasn't been disturbed for awhile.

✔ Check any area with recent water damage, termite damage, rot, or similar damage.

✔ Pound on hollow-core doors and double walls to disturb any insects inside. Place an ear (a stethoscope is better) against the wood and listen for rustling and scratching sounds produced by ant activity.

✔ Look for slitlike entrances or cracks leading to galleries (insect tunnels) and (though more likely in mainland areas) sawdustlike wood frass beneath the opening. This frass looks like tiny wood chips, not like the round pellets of drywood termites.

✔ Look for a nest in any objects that have hollows or spaces: cabinets, pianos, clocks, boxes, hollow-core doors, for example.

✔ If all else fails, you may have to trace the movements of the ants you see. Try putting out some sweets to lure enough workers out and see if they will lead you to the nest. This is real detective work!

Prevention of Carpenter Ant Invasions

The same techniques that are useful against termites, other ants, and cockroaches help keep out carpenter ants:

✔ Trees and bushes should be pruned so that they do not touch roofs, gutters, and sides of the house. These provide bridges for ants (and other pests) to enter the house.

✔ Keep yards clear of damaged and rotting trees, branches, and stumps and keep close watch on piles of lumber and debris that may hide wood-loving insects.

✔ Eliminate moisture sources. Any condition around the home that allows wood to become wet and remain damp (like a leaky faucet) invites wood rot and carpenter ants. Conditions allowing moisture should be corrected and damaged wood repaired. Gutters should be kept clear so that water does not overflow and damage wood below.

Control of Carpenter Ants

Outside nests are often found in areas high in moisture where there is lots of vegetation and rotted or windblown limbs and branches. Once you find the nest, destroy it and any stray ants. Pesticide formulations containing Dursban or diazinon can be used on nests outside (but do not use these indoors).

For nests inside walls, doors, and other objects, an aerosol such as Term-Out (Resmethrin) can be used. This has a convenient needle-nozzle on a flexible hose that allows spraying through small holes and cracks. Desiccating dusts, silica aerogels, and diatomaceous earth can be used, as described above for other ants, though these act much more slowly than insecticides. If you cannot eliminate the ants effectively, you may wish to contact a professional exterminator. For all chemicals, be sure to read and follow label directions.

BED BUGS
Order Heteroptera
Family Cimicidae
Cimex lectularius

Here Skugg lies snug
As a bug in a rug.
— BENJAMIN FRANKLIN

313 true bugs in Hawai'i, 214 native

"Nighty night, don't let the bed bugs bite," is a bedtime jingle still used by parents, though the jingle is from a time when bed bugs were more common. The bugs are still found worldwide, but they are not the regular bedmates they once were. Improved general sanitation seems to have reduced their numbers. We must, however, be watchful, lest they creep into our bedrooms even today.

Figure 10. Bed bugs have been associated with humans for a long time and are specialized to suck human blood. (Photo by Gordon Nishida)

Bed bugs (Figs. 10, 11) have been associated with humans for centuries and probably evolved along with prehistoric humans. Despite much research, there is still no evidence that they transmit disease organisms to humans. Though bed bugs are tolerated in many parts of the world, their bites and blood feeding can cause irritation and swelling in sensitive individuals, interrupt sleep patterns, and cause low hemoglobin count. They usually are associated with poor housecleaning and lack of sanitation. Don't let them share your bed with you.

Bed bugs move commonly from place to place on mattresses and bedding. They can be easily transferred on crowded public vehicles or easily introduced into the household in used mattresses and furniture, clothing, suitcases, and the like. The bug can also move on its own from house to house and from apartment to apartment through pipes and ducts, and cracks or holes in the walls.

Figure 11. Bed bugs feed while you are sleeping, painlessly puncturing your skin with their beaks and sneaking away with your blood. (Photo courtesy of Van Waters & Rogers Inc.)

Bed bugs are flattened (when hungry), wingless, reddish brown, oval insects, 1/8–1/4" long and about 1/8" wide. They have sucking mouthparts formed into a beak. After feeding, the bug elongates, fattens up, and turns dark red. A bed bug infestation can be recognized by the distinctive smell emitted from the bugs' stink glands; some describe the odor as that of rotting raspberries. Yellowish, red-brown droppings may also be seen on woodwork or walls and on bedding.

Bed bugs are found in cracks and crevices, in furniture, beneath loose wallpaper, in the frames of beds, in the folds of bedding, and in seams of mattresses. They emerge from their hiding places at night to feed on their sleeping human victims. The bugs occur in clusters and are most frequently found in bedrooms, in hotels, and in other public gathering places where sleeping victims are available for feeding. Bed bugs bite by inserting their beak into the skin and probing until they pierce a blood vessel. You don't feel the bite mainly because of the tiny size of the beak, hundreds of times smaller than a hypodermic needle. Bed bugs inject an anticlotting agent to keep the blood flowing as they feed. Both sexes suck blood. Females lay 10–50 eggs in batches in cracks and crevices on walls and floors near the bed; these are cemented to the surface and the shells remain even after hatching. Young hatch in about a week. The young nymphs feed daily, taking about 10–15 minutes to get their fill each time, and may feed 45 times before reaching adulthood. It takes about 2 months from egg to adult, and adults live 8–10 months. Adults withstand starvation well and may live a year or more without food under favorable conditions (yes, they can live longer without food than with it!).

Impact of Bed Bugs

A bed bug bite is usually painless, causing no feeling whatsoever. Most bites occur on exposed surfaces such as neck, arms, and legs, and usually while you are sleeping. If you suspect you are being bitten by bed bugs, set your alarm for 1:00 or 2:00 A.M., wake up, and investigate yourself and the bed with a flashlight. You may catch the little suckers at work! There are often multiple bites arranged in rows or clusters. Sensitive persons may react to the bite with slight swelling or inflammation. A small round or oblong bump with a tiny puncture at the center is common. If you are bitten, apply antiseptic to the bite. In case of severe bites or reactions, see a doctor.

Control of Bed Bugs

If you already have bed bugs, as a temporary measure you can use barriers to keep bugs from reaching you in bed. They do not fly but get onto beds when beds touch walls or by crawling up the legs of the bed. Try putting the legs of the bed in containers filled with soapy water; tuna or cat food cans will work. Or you can coat the legs of the bed with petroleum jelly for a few inches at the bottom. None of this will work, of course, if the bugs are already in bed with you! In some cases, the tricky buggers have been known to climb the ceiling and drop into bed with you.

All infested bedding should be washed. Infested beds and mattresses should be disposed of, fumigated, or treated with insecticides (outside the house). Be cautious when buying secondhand beds, mattresses, and other furniture. These should be inspected thoroughly for evidence of bugs before purchase and not bought if infested; after the furniture is home, even if there are no signs of infestation, it should be vacuumed, cleaned, and/or treated before use.

Bed bugs are controlled easily by pyrethrum sprays or dusts. You can also use a bulb duster to dust diatomaceous earth or sorptive dusts (see discussion under Dusts and Powders) into cracks and crevices in floors and walls. If the infestation is extremely heavy, you may wish to contact a licensed professional exterminator.

BEETLES
Order Coleoptera

1,984 species in Hawai'i, 1,367 native

Beetles are the largest group of insects. In fact, there are probably as many kinds of beetles as there are kinds of plants in the world. Better sanitation in food-processing plants, refrigeration, and the construction of furniture and fabrics from artificial products have reduced household problems with beetles. However, they occasionally become pests, particularly with household artifacts and handicrafts, baskets, furniture, and unprotected food products.

Beetles can be identified by the hardened first pair of wings that meet in a straight line down the back. They have chewing mouthparts, and the larvae do not look like the adults. Although some overlap occurs, beetle pests can be lumped into four areas: stored-food pests; wood and paper pests; fabric, carpet, and hide pests; and nuisances. The pest is usually the larval stage. Beetle larvae are whitish and grublike and are often difficult to tell apart from other larvae, such as moth caterpillars.

There is considerable overlap in the pest categories listed below. Many species do damage in all categories, but species are listed in the category in which they are most likely to be a pest.

Stored-Food Pests

Many insects discussed elsewhere in this book can become pests of stored foods. But certain beetles and moths are notorious stored-food pests in the home, as are some mites (see discussions in the chapters on Mites and Moths).

Most food products are contaminated by insects and insect parts to some degree. Most pests are imported into the house with foodstuffs already infested. If only the egg stage is present, it will not be obvious that the food is infested. Larvae hatch from eggs, feed and develop, and adults emerge in the product. If the infested food is not put into sealed containers after purchase, other food in the kitchen or pantry may become infested. Pests can bore through wrappers or cardboard containers, providing entry for other pests. Adults of some species feed in the same food that the eggs were laid in; others leave the food to pupate somewhere in cupboards or crevices in walls. Life cycles of stored-products pests vary, but management and control of these pests is about the same for all species.

Detection and Monitoring of Stored-Food Pests

Inspection and Proper Storage. Inspect packages before purchase. Inspect wrappers for damage or holes. Store foods in tightly sealed containers of a tough material so that adult beetles cannot bore through them. Glass or plastic containers with pressure seals are ideal. Screw-top jars are not bugproof; many insects can crawl down the threads and enter food unless there is a rubber gasket or other barrier at the top. If you find infested food after purchase, return it to the store if possible. Otherwise, it is probably best to dispose of the infested product. Be sure that the area where the food was stored is vacuumed and cleaned.

Storage at Low Humidity. Keep materials cool and dry to discourage pests. In Hawaiʻi, it is a good idea to store flour, cereals, dried fruits, and the like in the refrigerator.

Control of Stored-Food Pests

Freezing most foodstuffs for 4–5 days at 0°F will kill most stored-products pests. Most home freezers do not get that cold and so require a longer time (12–15 days) to kill pests. Some materials can be spread on a cookie sheet and baked for 30 minutes at 150°F. Lightly infested pet kibble can be salvaged this way.

Use of insecticides is not necessary nor recommended for use around foodstuffs. If the kitchen is heavily infested with adults flying around or larvae crawling on walls looking for places to pupate, apply silica aerogels or diatomaceous earth to cracks and crevices where pests have been noted. Take care to keep dishes and utensils off dusts that are applied. Pyrethrin sprays can also be used after removing contents of cupboards. Shelf paper can be used to separate dishes and utensils from insecticides, but shelf paper may provide hiding places for pests, including cockroaches.

Everything-but-the-Kitchen-Sink Beetles

The beetles listed below are often major household pests in Hawaiʻi, feeding on a wide variety of materials, even becoming pests of libraries and collections of artifacts or handicrafts.

✔ The **cigarette beetle** (*Lasioderma serricorne*, Family Anobiidae) (Figs. 12–14) is so named because it is a pest of tobacco and tobacco products. It

Figure 12. Cigarette beetle adults (in paprika) are tiny, reddish brown, and humpy looking because their heads are tucked down under their thoraces. (Photo courtesy of Van Waters & Rogers Inc.)

Figure 13. Cigarette beetle damage is often very obvious, because beetles bore through the food, creating a powdery, mildewy mess. Here, they've chewed through a plastic bag to get to tea and crackers. (Photo by Gordon Nishida)

actually has a large range of food preferences and will feed on spices, dried herbs, cereal products, nuts, seeds, cured fish and meats, upholstered furniture, hair, wool and other animal materials, cork, and other plant products (Fig. 13). In Hawai'i, it is often a pest of books, eating the binding and boring tiny round holes through the pages. The adult is very light brown in color, somewhat rounded in outline, and small, only about $1/16$–$1/10$" in length. The head is bent downward, giving the beetle a humped look. The larvae are yellowish white with light brown heads, C-shaped, very hairy little grubs. Egg to adult may take 4–6 weeks.

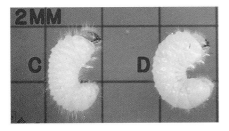

Figure 14. Grubs (larvae) of the beetle family Anobiidae are C-shaped and white or cream in color. The cigarette beetle larva is on the left, the drugstore beetle larva on the right. (Photo courtesy of Van Waters & Rogers Inc.)

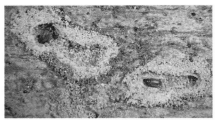

Figure 15. Adult drugstore beetles look very similar to cigarette beetles, but look slightly fuzzier. Here, adults emerge from pupae covered with fish food. (Photo courtesy of Van Waters & Rogers Inc.)

Figure 16. Wooden items can be damaged as drugstore beetles bore into wood to pupate and later emerge as adults. (Photo courtesy of Van Waters & Rogers Inc.)

✔ The **drugstore beetle** (*Stegobium paniceum,* Family Anobiidae) (Figs. 14–16) is very similar in size, shape, and color to the cigarette beetle and feeds on an even greater variety of plant and animal products (Fig. 16).

✔ The **herbarium beetle** (*Tricorynus herbarium,* Family Anobiidae) is very similar to the cigarette beetle in its tastes; it attacks spices, seeds, chocolate, furniture, leather, and also breeds in book bindings.

Grain and Flour Beetles

✔ The **sawtoothed grain beetle** (*Oryzaephilus surinamensis,* Family Silvanidae) (Figs. 17–19) attacks a wide range of stored-food products such as cereals, grains, flour, dried fruits, nuts, and seeds and also attacks candy, sugar, yeast, tobacco, and the like. The damage is mostly scarring and marring the surface of the food product. The adults are very small, reddish to dark brown, flattened, and thin, less than 1/8" long. The sides of the thorax have six "teeth" or spines on each side. The

Figure 17. The pattern of spines on the thorax of the adult saw-toothed grain beetle allows you to recognize this kitchen and pantry pest immediately. (Photo courtesy of Van Waters & Rogers Inc.)

Figure 18. The young of the saw-toothed grain beetle do not have spines on the thorax like the adults. (Photo courtesy of Van Waters & Rogers Inc.)

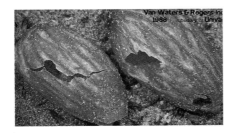

Figure 19. The saw-toothed grain beetle often damages products even inside sealed containers. Though they usually just scrape off the top of the food, they can—as in the almonds pictured—chew through and feed on the nutmeat, leaving the husk. (Photo courtesy of Van Waters & Rogers Inc.)

larvae are also slender, but whitish, with a pale head and tapering abdomen. The life cycle from egg to adult is 2–12 weeks; they may live up to 3 years.

✔ The **lesser grain borer** (*Rhyzopertha dominica*, Family Bostrichidae) (Figs. 20–21) feeds on the interior of nearly all grains, seeds, roots, cork, and drugs and also eats holes in wood and paper boxes. Both the adults and larvae cause feeding damage. The adults are brownish black, tubular, about $\frac{1}{8}$" long. The head is bent down rather than protruding forward like that of the sawtoothed grain beetle. The end of the abdomen is blunt rather than tapering.

Figure 20. The lesser grain borer is brown to black, with many tiny pits lining the wing covers. (Photo courtesy of Van Waters & Rogers Inc.)

Figure 21. Larvae of the lesser grain borer usually core out the inside of kernels of wheat, corn, and other grains, though they can be found in other foods also (L, larva; P, pupa). (Photo courtesy of Van Waters & Rogers Inc.)

Figure 22. Rice weevils are very similar to granary weevils (shown here), but have four lighter-colored patches on the back. Rice weevils and granary weevils are major pests of stored grain and are sometimes shipped out with the grain if storage silos are not properly monitored and treated. (Photo courtesy of Van Waters & Rogers Inc.)

✔ The **granary weevil** (*Sitophilus granarius*, Family Curculionidae) (Figs. 22, 23, 24) eats grain and grain products such as wheat, corn, oats, rye, millet, barley, macaroni, and others. The **rice weevil** (*Sitophilus oryzae*, Family Curculionidae) eats similar grains including rice. Adults have the characteristic weevil snout that extends downward under the head and may be as long as one-fourth the length of the body. They are both reddish brown and look similar, except that the rice weevil is smaller than the granary weevil (1/8" vs. 5/32") and has four lighter reddish yellow patches on the back. Egg to adult takes about a month. Adults may live from 7 months to over 2 years.

Figure 23. Granary weevils damage grain by eating out the kernels. (Photo courtesy of Van Waters & Rogers Inc.)

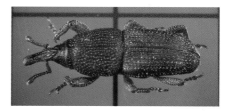

Figure 24. The granary weevil has the typical snout nose of the weevil family and is nondescript in color and shape. It is only occasionally found in Hawai'i because it prefers cooler climates. (Photo courtesy of Van Waters & Rogers Inc.)

✔ The **yellow mealworm** (*Tenebrio molitor*, Family Tenebrionidae) (Fig. 25) is often found in pet

Figure 25. The yellow mealworm beetle, shown here in oatmeal, is the largest pest you will find in your stored products. Shown are adults, pupa, and cast skins. (Photo courtesy of Van Waters & Rogers Inc.)

stores sold as food for lizards, geckos, and other animals. They occasionally are pests in stored products. They prefer grains that are damp or those that have not been disturbed for a long time. The larva is the familiar cylindrical mealworm that has alternating rings of yellow-orange and yellow-white. The adult beetle is black and somewhat shiny, flattened, and 1/2– 3/4" in length. Egg to adult takes from 40 weeks to nearly 2 years. Adults live 2–3 months.

Figure 26. The common name for this beetle, cadelle, comes from the French. The cadelle is found in packaged grains and grain products. (Photo courtesy of Van Waters & Rogers Inc.)

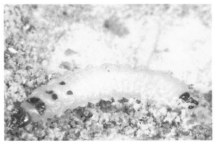

Figure 27. The larva of the cadelle has noticeable dark spots on the thorax. (Photo courtesy of Van Waters & Rogers Inc.)

✔ The **cadelle** (*Tenebroides mauritanicus*, Family Trogositidae) (Figs. 26–27) can be a serious pest of grain and grain products, with both adults and larvae causing damage. They also feed on nuts. The adults are shiny black to brown and about 5/16–1/2" in length. Larvae are slightly hairy, dirty yellowish white to grayish white, with black spots on the thorax and two dark, hooklike spines at the tail end. They can live 2–3 years.

✔ The **red flour beetle** (*Tribolium castaneum*, Family Tenebrionidae) (Figs. 28, 29) and the **confused flour beetle** (*Tribolium confusum*, Family Tenebrionidae) (Fig. 29) are similar in size, shape, color, and habits. They both reach about 1/7" in length and are reddish brown, flattened, and shiny. Both beetles feed on a wide range of grain and cereal

Figure 28. The red flour beetle adult shown here is similar to the confused flour beetle. (Photo courtesy of Van Waters & Rogers Inc.)

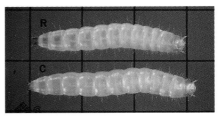

Figure 29. Larvae of the red flour beetle (top) and the confused flour beetle also look very similar. They are brownish yellow and somewhat flattened. (Photo courtesy of Van Waters & Rogers Inc.)

products, as well as seeds, peas and beans, roots, and dried fruits. Egg to adult takes about 2–4 months. They discolor and give a bad taste to the food they infest.

Dried Fruit and Sap Beetles

✔ The **dried fruit beetle** (*Carpophilus hemipterus*, Family Nitidulidae) and the **pineapple sap beetle** (*Urophorus humeralis*, Family Nitidulidae) (Figs. 30, 31) are unlikely to become pests in your home, unless you leave out dried or rotting fruit. They are mostly found outside and are pests of fresh ripe fruits and dried or drying fruits around the world. Occasionally, they attack drugs, bread, grain, and nuts, especially if these are moist or decaying. The beetles are small, about 1/8" long, black, with short, squared-off wings that do not cover the end of the abdomen. Both beetles have reddish brown to light-

Figure 30. Dried fruit and sap beetles belong to the family Nitidulidae and are easily recognized by the squared-off wing covers and clubbed antennae. Here, a sap beetle is shown on prickly pear cactus. (Photo courtesy of Van Waters & Rogers Inc.)

Figure 31. Larvae of dried fruit beetles bore into ripe fruit, causing the fruit to rot. They also feed on and foul dried fruits. (Photo courtesy of Van Waters & Rogers Inc.)

colored spots at the ends of the square wing covers. Egg to adult takes about a month.

If your fruit has been attacked by these beetles, it is probably a good idea to throw it out. Usually by the time beetles are seen, larvae have bored into the fruit and left a mess of droppings and larval and pupal skins. As they bore into the fruit, they introduce bacteria that often sour the fruit. That's why these beetles are also known as souring beetles.

Wood and Paper Pests

Wood-boring beetles can cause serious damage to wood structures and furniture, but damage occurs slowly over years, unlike subterranean termite infestations, which can do much damage in a short time. Once you spot the damage, you have time to identify the problem and decide what to do.

Prevention and Detection of Wood Borers

Most wood borers prefer to attack natural rather than finished wood. You can help prevent attacks by keeping your wood varnished, painted, or waxed.

Discovering beetle damage in wood does not necessarily mean that you have an active infestation. Exit holes and frass or powder indicate that beetles were there once, but they may not be now. The holes may have been there when the wood was installed or the furniture built. Signs that you have active insects are live adults, live grubs (larvae), or recent powdery frass. To test for frass that is recent, dust away frass already present. Check again in a few days; if you see new deposits of frass, insects are still working within the wood.

Control of Wood Borers

In structural wood, replace damaged wood. Electrogun treatments have also been used to kill beetles in paneling or structural wood (see discussion in the termite chapter). Spot treating with borax-based chemicals, pyrethrin/silica aerogels, or synthetic pyrethroids is often effective. Furniture beetles are unlikely to lay eggs in painted or varnished woods, so keep your wood well painted or varnished.

If you can be sure that an infestation is restricted to one item and if the item will not change character when heated, heating the material in an oven at 130–135°F for at least 6 hours can be effective. Be careful with heating books, because some paper ages more rapidly when heat is applied. An alternative is to freeze the material for about 15 days.

Infested furniture or wood can be kiln-dried or heated to 180°F for 30 minutes or frozen at 12°F for 15 days. If a kiln or large freezer is not available, or if the infestation is spread throughout the house, consult a pest control operator.

Powderpost Beetles

These insects are called powderpost beetles because of the fine, talcumlike dust they make while they are boring into wood. The dust often spills from the openings the beetles have bored and drops into small piles or patches on the floor underneath the item being eaten. This is a good sign of powderpost beetle infestation.

✔ The **powderpost beetle** (*Lyctus brunneus*, Family Lyctidae) (Figs. 32–34) is reddish brown, tube-shaped, and 1/16–3/16" long. Larvae are small, white grubs with an unusually large front end. Small round holes and fine dust appear on furniture, in hardwood floors, and in other wood. This beetle prefers oak, ash, hickory, maple, walnut, and bamboo.

Figure 32. Adult powderpost beetles are shiny black and tube-shaped. (Photo courtesy of Van Waters & Rogers Inc.)

✔ The **bamboo powderpost beetle** (*Dinoderus minutus*, Family Bostrichidae) is a cylindrical brown beetle about 1/8–5/32" long. The thorax is wider in front than behind. This beetle burrows into bamboo and likes rattan furniture. It also feeds on drugs, spices, grains, flour, and dried bananas.

✔ The **powderpost bostrichid** (*Amphicerus cornutus*, Family Bostrichidae) is a shiny black, tubular beetle, 5/16–9/16" long, with spines on the front part of the thorax. They usually are found outside in dead or dying twigs of many plants. They occasionally attack bamboo furniture.

Figure 33. Powderpost beetle larvae tunnel into wood, reducing it to fine dust. (Photo courtesy of Van Waters & Rogers Inc.)

Figure 34. Attacks by powderpost beetles leave tiny piles of very fine, talcumlike powder underneath the item attacked. (Photo courtesy of Van Waters & Rogers Inc.)

Fabric, Carpet, and Hide Pests

These beetles have wide-ranging tastes and feed on natural fabrics and carpets (cotton, wool) and other animal products such as feathers, leather, furs, rugs made of hides, hairbrushes, silk, horns, hair, tortoiseshell, and the like. They also are pantry pests, some feeding on cured meats and fish, beeswax, cheese, dried fruits and nuts. Some also attack the glue in book bindings. The damage is usually done by the larval stage. The adults often feed on pollen.

Most of the pests in this group belong to the family Dermestidae. They have conspicuous, brownish, bristly or hairy larvae. Shed skins look like larvae, but don't move. These beetles cause far more damage than clothes moths and eat a much broader range of materials. Large chew holes are usually concentrated in one area, in contrast to moth holes, which are scattered. Larvae wander about to pupate, so many are found on items they don't eat.

Prevention and Control of Fabric, Carpet, and Hide Beetles

Management of beetle pests is generally the same as for clothes moths:

✔ Destroy fabrics already damaged or thoroughly clean or freeze them.

✔ Clean fabrics and garments that have been stained; hang only clean garments in closets.

✔ Store susceptible, natural fabrics after cleaning in tightly sealed plastic bags.

✔ Avoid furnishings containing materials of animal origin.

✔ Use small area carpets that can be washed regularly, rather than wall-to-wall carpeting.

✔ Vacuum frequently.

✔ Store grains, cereals, and other foodstuffs in tight-fitting containers; discard infested materials.

✔ Remove old bird nests or other potential sources of infestation (beetles can develop by eating the feathers left in bird nests).

If chemicals are necessary, the same chemicals described for clothes moth control can be used. If you do not wish to deal with chemicals, consult a licensed pest-control operator. If the infestation is large enough or widespread, the house might have to be fumigated.

Hide and Carpet Beetles

Family Dermestidae

✔ The **furniture carpet beetle** (*Anthrenus flavipes*) (Figs. 35, 36) is blackish with patches of white, black, and brown outlined by yellow. It is very small, $1/16-1/8$" in length. The larvae are brown and yellow, with brown tufts of hair and three extra long tufts at the tail. This beetle was first noticed in furniture stuffed with horsehair, thus its name. Larvae feed on a wide range of animal products, including wool, hair, fur, feathers, horn, tortoiseshell, and silk. They also gnaw on cotton, rayon, leather, and soft wood. They live for 3–4 months.

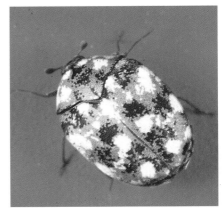

Figure 35. Carpet beetles are small and oval. The furniture carpet beetle is very colorfully dressed. (Photo courtesy of Van Waters & Rogers Inc.)

Figure 36. Carpet beetle larvae are hairy looking and feed on a wide variety of natural products. The furniture carpet beetle larva has dark brown hairs. (Photo courtesy of Van Waters & Rogers Inc.)

✔ The **carpet beetle** *(Anthrenus scrophulariae)* is also blackish speckled with white and has a band of orange-red scales down the center of the back. Adults are 1/16 – 1/8" long. The larvae are about 3/16" long and brown, with long hairs at the tail end. Larvae eat holes in fabrics and carpets. One way to tell that the damage is not caused by clothes moths is that there is no webbing associated with the damage. Besides carpets, the carpet beetle feeds on other animal products such as wool, feathers, leather, furs, natural-bristle hairbrushes, silk, and is a pest of museum specimens of plants and animals. Carpet beetles can also live off the feathers and insects in abandoned bird and wasp nests. The life cycle takes about 4 months; the adult probably lives for about a month.

Figure 37. The black carpet beetle is similar to the furniture carpet beetle, but is all black. The larva has lighter brown hairs covering its body. (Photo courtesy of Van Waters & Rogers Inc.)

✔ The **black carpet beetle** *(Attagenus unicolor* (Fig. 37); also the **wardrobe beetle,** *Attagenus fasciatus)* is dull black, with brownish legs and antennae. Adults are 1/8 – 5/32" long. Larvae may reach a length of nearly 1/2", are carrot-shaped, golden to chocolate brown, and have a tuft of long brown hairs at the tail end. Products damaged are similar to those for the carpet beetles listed above.

Figure 38. Adults of the hide beetle usually feed outside on the pollen of flowers. Larvae feed on and damage a variety of materials, but prefer protein, skins, and furs. (Photo courtesy of Van Waters & Rogers Inc.)

✔ The **black larder beetle** *(Dermestes ater)* is black, with yellowish gray, short, fuzzy hairs covering the body. It is about 5/16" long. The **hide beetle** *(Dermestes maculatus)* (Figs. 38, 39) is also black, with white hairs on the sides and underneath and is slightly larger at about 7/16" in length. Larvae are up to 3/4" long, black, and hairy. Both beetles infest a wide variety of food products, especially animal products such as feathers, horn, skins, hair, beeswax, dried fish, cured meats,

Figure 39. Hide beetles are important in the food web by helping break down carcasses of dead animals. Shown here are larvae and cast skins in a carcass. (Photo courtesy of Van Waters & Rogers Inc.)

dried cheese, and pure-bristle hairbrushes. They attack pet food in bags occasionally. They can also damage other materials, such as wood, cork, and the like by boring into them to pupate. The life cycle may take as little as two months to complete. Museums often use hide beetles to clean bits of meat off their skeletal specimens.

✔ The **cabinet beetle** *(Trogoderma anthrenoides)* is 1/16 – 3/16" long, black with a reddish pattern on the back. The larvae are yellow, with tufts of brown hair sticking out the end. They are pests of cereals and dried products like cocoa, soups, milk powders, grains, and seeds. They also attack dried fish, wool, feathers, hair, and skins.

Nuisances

Figure 40. This two-colored beetle *(Ananca bicolor)* often comes into homes at night, attracted by lights. (Photo by Gordon Nishida)

Figure 41. This beetle *(Thelyphassa apicata)* is attracted to light around houses and can cause skin reactions in sensitive individuals. (Photo by Gordon Nishida)

Though beetles are by far the largest group of insects, other than pantry pests and wood borers they do not affect us very much in our homes. Occasionally a stray beetle will wander in from the outside, possibly attracted to lights at night.

One beetle common at household lights is *Ananca bicolor* (Fig. 40). This beetle is long and slender, about ³/₈ – ½" long, with black wing covers, fairly long antennae, orange thorax, and brownish legs. This beetle is a member of the family Oedemeridae, the false blister beetles. Another member of the family, *Thelyphassa apicata* (Fig. 41), has caused blisters on people (see *What Bit Me?*) and also occasionally enters homes. *Thelyphassa apicata* is the same size and shape as *Ananca bicolor*, but is entirely brownish black.

We do not have the true blister beetles in Hawai'i, but the false blister beetles and possibly a few others can cause reactions in people with sensitive skin. The reactions are caused by oily fluids that are released by the beetles and used by them as protection from predators. Most people can handle the beetles with no reaction, but a few may be affected by these fluids and may get a rash or slight blistering.

Other beetles that often come to lights are click beetles (Elateridae) (Fig. 42). These typically long and narrow, often brownish, beetles do not put out an oily fluid. The outside trailing edges of their thorax often have sharp corners or spines. They protect themselves from predators by jamming a spine on the underside of the thorax into a groove and suddenly releasing it, flinging themselves into the air. The adults are harmless.

Other than the potential health problems, the beetles mentioned above will not damage anything inside the house. If you do not wish to handle them, a jar and an index card will come in handy, because the beetles do not fly readily. Place the jar over the beetle, slide the index card between the wall and the lip of the jar, cover the jar, and then either release or dispose of the beetle. A yellow bug light may help reduce the number of beetles attracted to your home.

Figure 42. Click beetles, such as *Simodactylus cinnamomeus*, are attracted to light around houses. (Photo by Gordon Nishida)

BOOKLICE
Order Psocoptera

135 species in Hawai'i, 89 native

Booklice (also known as psocids) are not related to lice, though they somewhat resemble head lice in shape. They are thought to feed on books, but, in fact, they are attracted to molds growing on the books. Outside species, called barklice (Fig. 43) live under bark, where they feed on mosses and lichens. Booklice (Fig. 44), *Liposcelis divinatorius*, do little damage, though they may damage books when they eat the starch in bindings. If you see groups of tiny insects in damp areas, suspect booklice.

Description: These are tiny, wingless animals, less than 1/16" long, usually with a head relatively large compared with the body, light-colored. Egg to adult takes slightly over 2 weeks; adults can live up to 6 months.

Evidence: Booklice are usually seen crawling on books or paper. They occasionally occur in foodstuffs, especially those with potential for growing

Figure 43. Hawai'i has a large number of native barklice. Adult barklice usually have wings. (Photo courtesy of Van Waters & Rogers Inc.)

molds, like grain. Booklice are also found crawling over moldy refrigerators, kitchen cabinets, and other such places.

Impact: These insects may do damage to paper or grain as they graze on molds. They can also damage books by eating the glue in the binding. They are mostly just a nuisance, but may indicate a high level of humidity (mold and fungi prefer higher relative humidity, 65% or above, so if booklice are present, it is an indication that your humidity is high).

Figure 44. Booklice are tiny, whitish animals most often seen crawling over pages of books or on paper. (Photo courtesy of Van Waters & Rogers Inc.)

Preferred Food: Molds and fungi.

Preferred Locations: Booklice are more frequently found in older wooden houses, warehouses, museums, and libraries, especially those that are not climate controlled.

Prevention: Reduce the humidity to control booklice. Keep potential food sources such as grains and bread in the refrigerator or in well-sealed containers.

Control: Booklice are usually not a serious problem, and chemical control is not recommended. Presence will probably increase or decrease depending on the weather. If occurring in foodstuffs, freeze the material or get rid of it. If on books, wipe off with a damp sponge; dry out in the sun; or fumigate in a plastic bag with a dehydrating agent such as anhydrous calcium carbonate (sold in hardware stores), diatomaceous earth, or silica aerogel. A dehumidifier is useful in extreme cases.

BUGS
Order Heteroptera

Figure 45. The bug *Oncocephalus pacificus* comes to lights to feed on other insects and sometimes bites people. (Photo by Gordon Nishida)

Most true bugs usually feed on plants, though some are predators of other insects. A very small minority affect humans—the bed bug, for example. Other bugs occasionally bite people, usually when they are mishandled. The most common biting bugs are covered in *What Bit Me?* Occasionally other bugs will take a test bite from a human. The reaction is normally a slight pain, though some bugs can deliver a wallop. The pain usually goes away after a short while.

One bug in Hawai'i, the Pacific kissing bug, *Oncocephalus pacificus* (Fig. 45), is sometimes found at lights and occasionally enters homes. It will bite. These bugs are not attracted directly by the lights, but feed on other insects attracted to lights. If these bugs become a problem, turn off all unnecessary lights and keep windows and doors and other openings well screened or closed. A yellow bug light may help reduce the number of insects attracted to outdoor lights.

CARPENTER BEES
Family Anthophoridae

His Labor is a Chant—
His Idleness—a Tune—
Oh, for a Bee's experience
Of Clovers, and of Noon!
—EMILY DICKINSON

73 bees in Hawai'i, 62 native

Sonoran Carpenter Bee
Xylocopa sonorina
(Figs. 46, 47)

Most people in Hawai'i recognize this large, beautiful black bee that is often mistaken for a bumble bee (*Bombus* spp.). Bumble bees do not occur in Hawai'i. Unlike honey bees, ants, and termites, which are social insects with a distinct caste structure, the Sonoran carpenter bee is a solitary bee.

Description: This is a large (about 1"), black bee (female) or yellowish brown or golden-colored bee (male) (Fig. 46). The females can sting,

Figure 46. The male carpenter bee is gold colored (*left*) and cannot sting; the female (*right*) is black and will sting if angered. (Photo courtesy of Van Waters & Rogers Inc.)

Figure 47. Carpenter bees bore holes in wood to make their nests. (Photo courtesy of Van Waters & Rogers Inc.)

although they do not usually do so unless they are angered; the males do not have a stinger and cannot sting.

Evidence: These large bees zoom swiftly around buildings, making loud buzzing noises. They may appear frightening because of their size and apparent disregard for humans as they seem to blunder around. They make large (up to 1/2"), round entrance holes in wood trim, siding, or other wood structures outside; loud buzzing may come from inside the wood.

Impact: Females are often seen visiting flowers, gathering nectar and pollen to feed to their young; they are very important as pollinators. Carpenter bees are active year round in Hawai'i. They prefer nesting in soft woods (redwood, pine, canec) and in forests nest in dead limbs of trees. Unfortunately, they sometimes take wooden structures for giant dead trees and bore into beams to make their nests (Fig. 47). The female chews into the wood with her powerful jaws and makes a tunnel in which to lay eggs. Over time and reuse by other bees, a small maze of branching tunnels may develop. The female enters the wood against the grain, and then after about 1" of chewing, turns suddenly along the grain of the wood to make the main part of the tunnel. An egg is placed in the far end of the tunnel, and the cell is provisioned with a mixture of honey and pollen for the young to feed on after the egg hatches. She then seals the cell with chewed wood and saliva and lays another egg. There may be several eggs in cells along the tunnel. Eggs hatch in 2–3 days, and the cycle from egg to adult takes 1–3 months.

The sting of a female carpenter bee can be very painful, but is usually not dangerous. The sting wound can take a long time to heal; if a bad reaction occurs to the sting or if you are sensitive to other insect venoms, see a doctor immediately.

Preferred Food: Carpenter bee adults feed on plant nectar; larvae are fed pollen and nectar. Adults do not eat wood, but bore into it to make nests.

Preferred Locations: These bees are usually found in forests, but they occasionally tunnel into wooden structures. Structures commonly attacked are roof trim and siding, wooden roof shingles, eaves, doors, windowsills, canec ceilings, telephone poles, fences and wooden outdoor furniture, decks, and railings.

Prevention: Unpainted wood is preferred over painted wood by bees, so paint or varnish exposed wood regularly. Use wood that is pretreated. Avoid using their preferred soft woods in areas where carpenter bees are active.

Control: Pyrethrum-based sprays can be used to treat the holes. Read the insecticide label for appropriate chemicals and for application instructions. Some pyrethrins are formulated with silica aerogels, which leave a residual dust that will make the treated areas unattractive to females. The best time to treat is in the evening after sundown when bees are in and less active. Simply plugging holes usually will not work—bees will chew right through most caulk or wood fillers in a day or so. You might have success with filling holes with steel wool covered by stapled-on metal screening after treatment. Another method is to plug holes with nylon stocking or other porous material and saturate the material with a poison. You will probably hear loud, angry buzzing coming from behind your plug. Be careful—you've made the bees mad! Unfortunately, these methods leave dead bees and young inside the tunnels, to become food for something else.

If you do not wish to use chemicals, you can wait until the bees leave the nest and then remove and replace the wood that has been tunneled with treated wood.

CENTIPEDES
Class Chilopoda

A centipede was happy, quite,
Until a toad in fun
Said, "Pray, which leg moves after which?"
Which raised her doubts to such a pitch,
She fell exhausted in the ditch,
Not knowing how to run.
— DITTY

24 species in Hawai'i, 12 native

Most people try to avoid these many-legged creatures at all costs. But centipedes are really not as dangerous as they look. The bite is sometimes extremely painful, but not deadly. Centipedes can be distinguished from millipedes by counting the number of legs on each body segment: centipedes only have one pair of legs per segment; millipedes have two pairs. Centipedes tend to be flattened; millipedes are cylindrical or tubular.

Large Centipede
Scolopendra subspinipes
(Fig. 48)

Description: This is a large (up to 9" long), reddish brown, multilegged insect relative. "Centipede" means "100-legged," but adults actually have only 22 pairs of legs, one pair per body segment. The young are often bright blue or green and orange. Adults and young can move very quickly, propelling themselves forward with a wriggling, snakelike motion.

Figure 48. Centipedes unnerve most people. However, they probably do more good than harm, being excellent predators. (Photo by Gordon Nishida)

Evidence: When in the house, centipedes are usually seen crawling across the floor; occasionally they climb walls.

Impact: Centipedes "bite" with the tips of the enlarged first pair of legs, which carry poison glands. Though they are longer than the rest and look menacing, the last pair of legs do not sting. Centipedes bite humans in self-defense, as when stepped on, grabbed, or squeezed while they are hiding in clothes or bedding. *Scolopendra*'s bite may cause intense pain followed by swelling and reddening around the wound lasting a few hours. The wound is slow to heal and may become infected. The skin surrounding the wound may eventually die and slough off. Treatment other than antiseptic is generally not necessary, though an ice pack applied to the wound may ease the pain and help reduce the swelling. If symptoms persist, see a doctor.

Preferred Food: Centipedes hunt insects, worms, and slugs and are especially fond of cockroaches. They are excellent predators and can be useful in keeping populations of pest arthropods in check.

Preferred Locations: Outside, centipedes hide under rocks, logs, boards, piles of trash or mulch during the day and come out at night to hunt prey. They have a narrow comfort range and sometimes enter the home to escape from too much water in cases of heavy rains or to find a damper place in dry weather. They sometimes also enter homes to hunt for prey or, in some cases, to escape insecticides sprayed where they ordinarily hide.

Prevention: The best preventive measure is to remove hiding places on the ground such as rocks, loose boards, piles of trash, and ground covers. By cleaning up trash and clearing out places they like to hide or hunt, you can make it less likely they'll hang around. Get rid of cockroaches and other potential food for centipedes to keep them out. Another tack is to clear a zone around the house to discourage centipedes from wandering into the house. Watering of lawns and plants may make it nice and friendly for the centipedes, so you may want to modify your watering schedule or change (or remove) your ground cover. In exceptional situations, a perimeter insecticidal spray might be tried, but remember that insecticides often kill centipedes slowly and might force them from their outside hidey-holes into the house. Some houses seem to have been built over centipede "cities" and have a greater than normal rate of visitation by these many-legged creatures. If you happen to live in such an area and do not want to put up with them, you must change the environment and chase out the centipedes by either removing their food or making their hiding and breeding places uncomfortable or unlivable. This could require a major relandscaping effort.

Control: Prevention is the best control. Insecticides work, but somewhat slowly. Stomping centipedes is effective, but it leaves a squishy mess; do not stomp them barefooted, because the "fangs" could penetrate your foot.

COCKROACHES
Order Dictyoptera

Long after the bomb falls
and you and your good deeds are gone,
cockroaches will still be here,
prowling the streets like armored cars.
— TAMA JANOWITZ

19 species in Hawai'i, 0 native

Cockroaches must be one of the most hated creatures known to man. Their greasy, bristly appearance, filthy habits, and potential disease-spreading capabilities raise the meaning of repulsiveness to new levels. Cockroaches feed on almost anything edible and even sample things that are not edible. They foul human foodstuffs with their vomit and excrement. Their pheromones leave a strong, foul odor on materials they visit. Is it any wonder we find them unwelcome house guests?

Life histories of all cockroaches are similar. Eggs are laid in a beanlike or podlike capsule called an ootheca (Fig. 49). In some species these "egg cases" can be seen sticking out of the end of the abdomen as the female cockroach moves around. The young, or nymphs, look like smaller versions of adults but are wingless. These young develop in 2–18 months to full-grown adults.

Cockroaches have long antennae, flattened oval bodies, and spiny legs. The front wings are partly hardened (though not as hard as those of beetles), and the hind wings are clear. All are very active, fast-running insects. Four species occur regularly in houses in Hawai'i; a fifth might wander into the house occasionally. These are described below.

Cockroaches are active at night. If you turn on a light, you can spot them scurrying for cover. They leave black droppings (feces) in their hiding places and drop or glue their egg cases in various places. You may also see numerous tiny, young nymphs that have just hatched from egg cases. Cockroaches clump together in favorable hiding places, partially because of an aggregation pheromone some release. This is pungent and foul-smelling. The pheromone is also present in feces, so cockroaches tend to come to

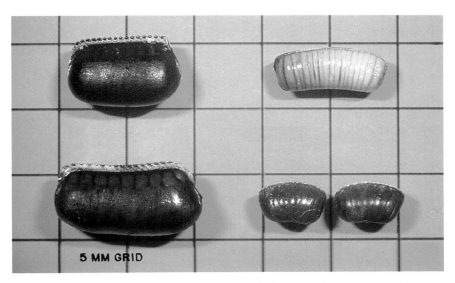

Figure 49. Cockroach egg cases have the same basic design, with many eggs inside a pod or capsule. The egg cases are different enough that you can usually tell what species you have by looking at the egg capsule. (Photo courtesy of Van Waters & Rogers Inc.)

sites previously visited by other cockroaches. A house or apartment heavily infested with cockroaches is likely to smell bad.

Cockroaches like to gather in cracks and crevices, in cupboards, under shelf paper, under refrigerators, under burner areas of stoves—in short, almost any place that is dark, moist, warm, and near food. They migrate from buildings and through apartments via elevator shafts, drains, electrical conduits, ducts, vents, doors, windows, and similar structures.

Impact of Cockroaches

In homes, cockroaches feed on many items, including bits of food, leather, hair, paper, starchy materials, wallpaper, books, feces, blood, and even toenails. They also feed on garbage and sewage. Actually, they will feed on or sample almost anything. They have even been seen feeding on termites during swarms.

Research has not yet proven that cockroaches are actually involved in the natural transmission of disease organisms (that is, that they infect people and people get sick). However, many pathogenic organisms have been taken from their gut and feces, including those causing gastrointestinal diseases like salmonella and dysentery, skin diseases, and diseases like typhus

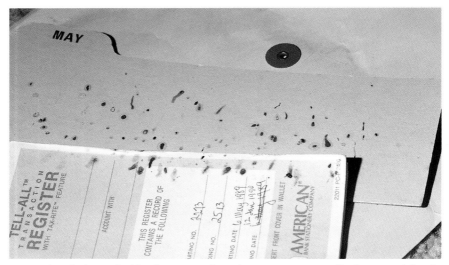

Figure 50. Black streaks like these on surfaces (paper, counters, walls, for example) indicate the presence of cockroaches. These streaks were probably made by a German cockroach. American and Australian cockroach vomit streaks are much larger. (Photo by Gordon Nishida)

and polio. Cockroaches wander through all kinds of organic wastes, picking up bacteria and other microorganisms. They then crawl over your kitchen counters, food, dishes, and utensils and taint them with their vomit (Fig. 50) and feces. We find it difficult to draw the conclusion that these creatures are not potentially harmful to humans.

Cockroach bites may occur where cockroaches are uncontrolled and have built up heavy populations. They may gnaw toenails and fingernails of a sleeping, sick, or helpless person and bite into the surrounding flesh. Reports of bites are rare when good hygiene and cockroach control are practiced.

Some research shows that cockroaches may be as important as house dust mites in triggering allergic reactions. Read the section on house dust mites for further information on these allergies.

Prevention of Cockroach Infestations

Super-cleanliness is important in keeping down infestations or preventing reinfestation. However, a totally sterile environment would be necessary to discourage cockroaches completely. This is impossible, especially in a household with children or pets. In apartments and condominiums, where cockroaches can travel easily throughout the structures, it is even more difficult to keep your own place cockroach-free. Often, the best you can hope for is to keep populations low enough so that you do not see cockroaches in the daytime. If populations are so high that some cockroaches are forced to be out during the day, it is time to take serious action.

Detection and Monitoring of Cockroaches

For every cockroach you see in your kitchen, there are probably 50 others lurking in their hideouts. To find out the level of your cockroach population or to see if control measures have reduced populations, sticky traps can be used. Also, the German cockroach, for example, tends to hang around areas with good hiding places and nearby food. You can figure out where these areas are by placing sticky traps around the kitchen.

Sticky traps without poison (some have attractant baits) are good monitoring devices (see the discussion on page 8). Use as many traps as you can afford in the kitchen and put some also in the bathroom. Place traps along cockroach paths and gathering places; do not put traps in open areas. Think of dark, warm, moist areas like cracks and crevices, in drawers and cupboards, near plumbing fixtures under sinks, along floor molding, where countertops meet walls, and under refrigerators, for example. Obviously, you can also use these traps to reduce cockroach populations, but usually they will not be enough to control your populations completely.

Control of Cockroaches

Someone once said that a cockroach is so ugly that the best way to control it is to put a mirror in front of it; this will so frighten the cockroach that it will scurry away. If only it were that simple! Unfortunately, it takes a powerful effort to manage good cockroach control. But, it can be done.

Physical Control Measures

Many of the following anticockroach measures have been discussed in previous sections. However, they are important to cockroach control and are repeated here.

Sanitation and Proper Storage of Food. This is the easiest and most obvious means of control. It reduces cockroach populations by reducing available food. Food should be stored in the refrigerator or in tightly sealed containers. After meals, clean and wipe up all crumbs on counters and floors. Do not allow uneaten pet food, canned or dry, to remain out on the floor (this also attracts ants). Wash dishes immediately, or rinse off or soak in soapy water until you have time to wash them. Do a periodic complete cleaning of the kitchen, focusing on countertops and areas where grease collects: drains, stoves, ovens, grout on ceramic tiles, hood over the stove, for example. Clean under all small appliances: microwave, coffeemaker, toaster oven, and the rest. Store trash and garbage in tightly sealed containers or empty frequently.

Preventing Entry/Modifying Habitats. Screen vents and windows and ensure that screens, windows, and doors fit properly. Screen, caulk, or seal areas around plumbing where cockroaches may enter through floors and walls. Repair any leaks in plumbing fixtures, because cockroaches are attracted by the moisture. Where possible, reduce cockroach hiding places by caulking or painting cracks and crevices. Caulking cracks is a good idea but may be impractical in older dwellings because nymphs can squeeze through very tiny gaps (about 1/25").

Vacuuming. Use the vacuum liberally in kitchens to vacuum out-of-the-way areas. You can use a strong vacuum to suck up live cockroaches and egg cases from hideaways. After cleaning, put the vacuum bag in a sealed plastic bag and dispose of it.

Chemical Control

It is best to try permanent modifications and sanitation measures, as described above, first; then apply insecticidal measures when needed. Insecticidal sprays may not work because cockroaches have developed resistance to many of them. Anyway, they are potentially hazardous and temporary measures at best.

Dusts like boric acid, diatomaceous earth, and silica aerogels are effective against cockroaches (see discussions of Dusts and Powders under Chemical Control), and if used properly, can provide long-term control safely and inexpensively.

Boric acid is an old remedy and still effective because cockroaches have not developed resistance to it. It is slower to work (5–10 days to have an effect on the population), but it is relatively nontoxic. Blow it with a bulb duster into cracks and crevices, behind walls, under stoves and refrigerators, and other dark places. Dust it evenly over a wide area to maximize the chances a cockroach will walk over it; they will eat it because they are frequently grooming themselves. Do not dust areas where children might enter and ingest the dust. Avoid placing dishes and utensils directly over the dust. Shelf paper will protect dishes from the dust, but it also provides hiding places for cockroaches and a nice place to leave egg cases. Silica gels can be applied in the same manner as boric acid. In Hawaiʻi, these dusts should be vacuumed up and replaced periodically because of their tendency to pick up moisture and clump.

Bait stations (see discussion of Baits under Chemical Control) that contain hydramethylnon (such as Combat) provide a quicker kill. Cockroaches are attracted to the bait, feed on the poison, and die usually within 2–4 days. Hydramethylnon bait stations are reportedly very safe, because they have relatively small amounts of the chemical. It is said that even if a child could swallow the contents of the black bait stations, he/she would have to eat the contents of 40 or more to get sick. However, even at sale prices, these bait stations are much more expensive than boric acid. But using bait stations in combination with boric acid dusting will be extremely effective.

In severe cases, professional exterminators may have to be called. They can eliminate most of the cockroaches and allow you to begin a proper control program when populations are low.

Proper Use of Bait Stations for Cockroaches

✔ Don't leave any food out. Clean up well before setting out bait stations. Bait will work only if there is no other food competing with it. Don't spray anywhere near stations, because they will be contaminated and the cockroaches will avoid the area.

✔ Use at least the number of bait stations suggested in the product directions. In heavy infestations, use as many as you can afford in kitchens and bathrooms.

✔ Place stations under and around sinks, under stove and refrigerator, on countertops, in cupboards and drawers, near garbage cans, behind toilets, and near showers and bathtubs.

✔ Place stations in corners or on surfaces where walls, floors, cabinets, and shelf surfaces meet. Do not place in open spaces, because cockroach routes are close to walls or appliances.

✔ The manufacturer's label directions provide a time period for replacement of bait stations. This is an estimate of the time it takes for cockroaches to eat all the bait. This depends on how many cockroaches you have, so a better plan is to change bait immediately when you start to see cockroaches again. This usually means that the bait is all eaten.

American Cockroach
Periplaneta americana
Family Blattidae
(Fig. 51)
Also known as Water Bug, Palmetto Bug

The American cockroach is large, about 1½" long, reddish brown, and its thorax is vaguely outlined in a cream color. It is frequently found in commercial establishments, such as restaurants, grocery stores, warehouses, and such. It is usually not as common in the house as the smaller cockroaches, although in some homes in Hawai'i it is the most commonly seen cockroach. Egg cases are carried up to 6 days. Each has about 15 eggs; the cases are brown when laid and turn black in a few days. This cockroach likes to glue its dark brown egg cases to surfaces (preferably wood), especially in corners of cabinets, underneath counters, and in drawers. From egg to adult takes about 280–620 days. Where this cockroach has been feeding, you may find large blackish brown streaks of dried vomit.

Figure 51. The American cockroach is reddish brown with a poorly outlined thorax and no slashes at the base of the wing covers. This female's egg case protrudes from the rear. (Photo courtesy of Van Waters & Rogers Inc.)

Australian Cockroach
Periplaneta australasiae
Family Blattidae
(Fig. 52)

The Australian cockroach is similar to the American cockroach in size and appearance, but is slightly smaller and has a sharply defined whitish ring around the thorax and whitish slashes at the base of the wings. This cockroach is more often found outside, but occasionally invades houses. It also flies readily.

Figure 52. The Australian cockroach is slightly smaller than the American cockroach. Its thorax is clearly outlined in white and it has white slashes at the base of the first pair of wings. (Photo courtesy of Van Waters & Rogers Inc.)

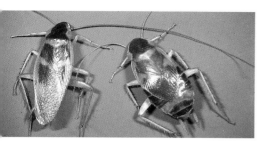

Figure 53. The brownbanded cockroach is about the same size as the German cockroach, but has a couple of wide, brown bands that cross the middle of the back. The female has shortened wings. (Photo courtesy of Van Waters & Rogers Inc.)

Brownbanded Cockroach
Supella longipalpa
Family Blattellidae
(Fig. 53)

The brownbanded cockroach is about ½" long and has two light brown crossbands on its back, beginning at the base of the wings and alternating with dark bands. The females have shorter wings that do not extend to the end of the body. This species is not as common as the German cockroach and is usually found in higher and drier locations, such as in picture frames, cupboards, television sets, and radios. It likes to glue its egg cases on walls and ceilings, especially in cracks and crevices and behind paintings and other objects on walls. The egg case usually has about 16 eggs; egg to adult takes 95–280 days.

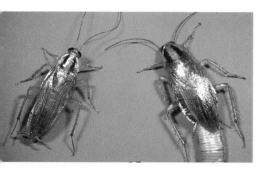

Figure 54. The German cockroach can be identified by the two vertical black stripes that run down the thorax. (Photo courtesy of Van Waters & Rogers Inc.)

German Cockroach
Blattella germanica
Family Blattellidae
(Fig. 54)

This is the most common and obnoxious cockroach in Hawai'i and the mainland United States. It likes high temperatures and does well in hot climates.

The German cockroach is about ½" long and has two longitudinal black stripes on the thorax. The nymphs may have the bands extending down the abdomen. This species prefers kitchens or bathrooms and loves warm places, such as under the refrigerator or next to the water heater. Though you may not see cockroaches, finding the light brown egg cases is evidence of their presence. The female carries the egg case extending from the end of her abdomen until about 12 days before the eggs hatch; then she drops it anywhere. Up to 48 eggs may develop in each egg case; the female produces four to five egg cases in her lifetime. Adults may live up to a year.

Figure 55. The burrowing cockroach looks like a beetle and is usually found in or on soil; it likes to burrow underneath objects. This female is carrying her egg case. (Photo courtesy of Van Waters & Rogers Inc.)

These cockroaches leave smaller streaks of vomit than do American cockroaches and also drop fecal pellets. The fecal pellets contain an aggregation pheromone, or scent, that attracts other German cockroaches.

Burrowing Cockroach
Pycnoscelus indicus
Family Blaberidae
(Fig. 55)

This cockroach was previously known as the Surinam cockroach (*Pycnoscelus surinamensis*). It is dark-colored, almost black, with the thorax darker than the brownish wings. It is shiny and somewhat beetlelike.

Typically, the burrowing cockroach is found outdoors in loose soil or under objects on the ground. It damages underground parts of some plants, but is usually a scavenger. This species is often abundant on poultry farms and is an intermediate host of the poultry eye worm. The cockroach is eaten by the toad *Bufo marinus* and, at times, provides up to 50% of the toad's diet. Sometimes this cockroach strays into the house and occasionally becomes a pest of houseplants, especially on the lānai.

CRUSTACEANS
Class Crustacea

This class includes crabs, crayfish, lobsters, and similar arthropods, most of which are aquatic. Some forms are terrestrial and are sometimes found in large numbers in damp places, including bathrooms and basements in the house.

Amphipods
Platorchestia platensis
Subclass Amphipoda: 9 terrestrial species in Hawai'i, 6 native

Amphipods (Fig. 56) are small shrimplike creatures that occasionally enter the lower floors of homes. They are less than ¼" long, brownish to pale in color, changing to orange when dead. Signs are little orange bodies strewn about the floor or funny-looking shrimp that jump. When present in large numbers, they may present a cleaning problem, but that's about all.

Amphipods feed on organic material. They are usually found in moister parts of the Islands and occasionally enter houses seeking water. They have a great ability to sense moisture and, for example, will jump onto coffee-table tops if a wet glass has left some condensation. In the house they are not much of a problem; if pesky, seal entryways and eliminate sources of standing moisture. If necessary, dehydrators such as silica aerogels or abraders like diatomaceous earth should work against amphipods.

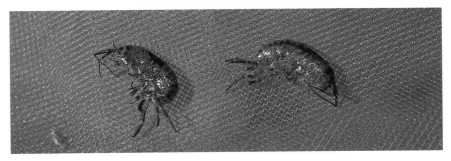

Figure 56. Amphipods need lots of moisture to survive and are only found in very moist situations. They are shrimplike and often turn orange after death. (Photo by Gordon Nishida)

Figure 57. Sowbugs are many-legged crustaceans that look like pillbugs but cannot roll into a ball as pillbugs can. (Photo by Gordon Nishida)

Sowbugs and Pillbugs
Subclass Isopoda: 46 terrestrial species in Hawai'i, 17 native
Porcellio laevis (Sowbug)
Metoponorthus pruinosus (Sowbug)
Venezillo parvus (Pillbug)

Sowbugs (Fig. 57) and pillbugs are the familiar grayish, sometimes brownish, armadillo-like isopods that are often quite common in damp places: under stones, logs, bark of trees, and in other places with damp, decaying materials. They usually feed on rotting vegetable matter, but have been known to damage live plants. Pillbugs can roll themselves into a ball; sowbugs can't. They sometimes enter homes and are usually in the damper parts of the household. They do not bite or sting or usually cause any damage, but can be nuisances if they enter the house in large numbers. After they die, they often turn whitish if left long enough.

Normally no control is necessary; isopods will disappear as conditions dry. If you have a recurrent problem, you may wish to change the conditions that attract these armored animals. Remove sources of moisture such as leaking pipes and remove any organic debris, including matted leaves, cardboard boxes, stacks of paper, and other items that may pick up and hold moisture. In exceptional cases, leaves and heavily watered ground cover outside the house may have to be removed.

EARWIGS
Order Dermaptera

24 species in Hawai'i, 10 native

Earwigs are an ancient group of insects. Their odd name comes from an old superstition that they crawled into or attacked people's ears and bored into their brains. Earwigs have a pair of pincers (or forceps) at the end of the abdomen that they use to grab and hold their prey and to defend themselves against enemies. They also use the forceps to fold their wings back under the tiny, squarish wing covers on their back. Males have larger forceps than females. These insects look frightening to people because of the sometimes-large pincers, but earwigs are not a menace to humans. They are mainly scavengers, but also feed on other insects; some become pests in households by feeding on grains and other starchy things. However, earwigs are now known to be largely beneficial, foraging at night on eggs, young, and adults of smaller insects.

Black Earwig
Chelisoches morio
Ringlegged Earwig
Euborellia annulipes
European Earwig
Forficula auricularia
(Fig. 58)
Striped Earwig
Labidura riparia

Description: These are medium-sized insects, usually brownish or yellowish, except for *Chelisoches morio*, which is entirely black except for whitish rings around the lower part of its antennae.

Evidence: Earwigs are active mostly at night, hiding in cracks and crevices during the day.

Impact: There is usually not much impact on humans, because they feed on organic debris and other insects. Occasionally, they may attack starchy sub-

Figure 58. Earwigs use their pincers to fold their wings under the tiny wing covers (tan colored in the European earwig shown here) on their back. The pincers are also used to grab prey. (Photo courtesy of Van Waters & Rogers Inc.)

stances such as grain. They have the potential of biting, although the bite is not serious. The pincers are surprisingly strong and can give you a sharp nip, though they are not strong enough to draw blood.

Preferred Food: Varied. Some species prefer plant material, others prefer catching insects, and still others like to feed on decaying matter.

Preferred Locations: Some prefer bases of plants; others like hiding in organic debris. Some may enter homes and hide under rugs, cardboard boxes, or other items on the floor. They like moist areas, so laundry rooms and bathrooms are the rooms most often invaded.

Prevention: If your house is at ground level, clear a zone several feet wide around the home of nonessential plants, plant debris, mulch, boards, and other hiding places. The object is to create a zone that sunlight will dry out and make disagreeable to earwigs. Prune away foliage that touches the house.

Control: Try to resist the urge to kill these insects, even though they may look scary. If large numbers are around and you can't stand them, insecticides will work well; be careful with their application. Powders such as boric acid, diatomaceous earth, or silica aerogels should also work, but these may not be very effective because earwigs are usually found in damper areas of the house. The physical control of earwigs must be combined with cleaning up their hiding places or they will likely return.

FLEAS
Order Siphonaptera

Great fleas have little fleas upon their backs to bite 'em,
And little fleas have lesser fleas, and so ad infinitum.
And the great fleas themselves, in turn, have greater fleas to go on,
While these again have greater still, and greater still, and so on.
— DE MORGAN

11 species in Hawai'i, one native

Fleas are wingless, leathery-bodied parasites of warm-blooded animals. They are closely related to flies. In early human evolution, humans were infested with fleas as our domestic animals are today. We were plagued primarily by the human flea, which also infested other animals. Modern sanitation has just about wiped out problems with this flea on humans. Unfortunately, the cat flea—found on both cats and dogs in Hawai'i—causes us almost as much misery as other major pests such as cockroaches and termites.

Fleas are powerful jumpers, able to jump up to 150 times their length. The tough shell is resistant to scratching, the side-to-side flattening allows fleas to skitter through hairs rapidly and makes them hard to catch with pet's teeth or owner's fingers, and the bristly hairs allow fleas to cling tightly to animal fur.

Fleas probably arrived in Hawai'i with European or American ships sometime before 1809. The Hawaiian name for flea, *'uku lele,* means jumping louse or, perhaps more broadly, jumping bug. That name was adopted for the small four-string guitar that is so often identified with Hawai'i. The most accepted version of the origin of the name of the *'ukulele* is that the rapid, jumping movements of the fingers in playing the instrument resemble the movements of fleas.

The only flea species you are likely to see in your house in Hawai'i is the cat flea—yes, even on dogs. The human flea, *Pulex irritans,* is found occasionally in Hawai'i, but is rarely a problem. The oriental rat flea, *Xenopsylla cheopis,* is present on all major islands in Hawai'i. This flea, a parasite of rats, may bite humans when there are a great many infested rats, or when rats have left their nests.

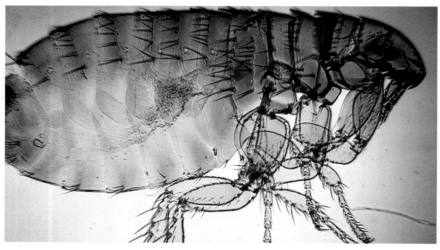

Figure 59. Cat fleas are the fleas that usually infest pets and bite people in Hawai'i. They are incredible jumpers and their bodies are covered with tiny, backward-pointing spines. (Photo courtesy of Van Waters & Rogers Inc.)

Cat Flea
Ctenocephalides felis
(Figs. 59, 60)

The cat flea is the flea most commonly found around homes and is often reported biting humans; it is found on all major islands. Fleas are reddish brown, wingless, leathery bodied, and flattened from side to side. The females are about 1/16" long, expanding up to 1/5" after a blood meal; males are smaller.

Both males and females feed on mammals by piercing the skin with their mouthparts and sucking blood. Flea eggs, laid one at a time on the host, drop to the floor, hatch, and develop into wormlike larvae. The black specks that you see around flea bites and in pet bedding are not eggs, but flea feces mixed with dried

Figure 60. Fleas are flattened from side to side, so they can easily slip through even very thick fur. Here, a cat flea defecates on human skin. (Photo courtesy of Van Waters & Rogers Inc.)

blood; if you wet the feces, they will turn reddish. Larvae live where eggs have dropped, not on the pet. Larvae feed on dried blood particles, excrement, and other organic debris in cracks, crevices, carpets, and animal bedding. The larva eventually spins a silken cocoon and pupates. The adult develops in 1–2 weeks, but requires some stimulus such as vibration to escape from the cocoon. It can survive from 1 to 5 years in the cocoon if there is no suitable stimulus. When the adult emerges, it is hungry for blood and looks for a host to jump onto. It can go for a few months without blood and wait several months for another meal once it has fed. Fleas prefer dogs or cats but will bite humans when the fleas are overly abundant or when the host pets have died or are no longer present.

Impact of Fleas

Flea bites are sometimes numerous on certain individuals, and allergies can develop in sensitive humans. Allergies in pets may lead to hair loss, heavy scratching, and red, raw, sometimes infected skin. Some people are more attractive to fleas than others. Fleas are attracted by various skin secretions and by carbon dioxide emissions. Differences in immune systems may have something to do with differences in attractiveness of humans.

Fleas are known to transmit diseases, such as typhus and bubonic plague. The oriental rat flea is usually the major carrier of these diseases, although other fleas may carry them also. Plague has been present in Hawai'i since the middle 1800s, but there has been no outbreak of the disease here since the late 1950s.

Dog owners in Hawai'i should know that the cat flea is host of the dog tapeworm (*Dipylidium caninum*), a parasite that lives in the intestines of dogs. If your dog has fleas, it probably has tapeworms. To be infested with tapes, the dog must eat the flea; this may happen when the dog is biting itself to relieve the itch of the biting fleas. Signs of tapes are obvious. The worm sheds segments filled with eggs, and these segments pass in the feces. You can see these segments on the anus or in fresh feces; when fresh, they are white and wriggle like worms. Your veterinarian can confirm the presence of tapeworms.

The dog tapeworm can live with minimal harm to your dog if the dog is otherwise healthy. However, it is best to have your dog wormed occasionally by a veterinarian. If you continue to have flea problems, however, your dog will be reinfested, so ultimately you must tackle the flea problem to keep your dog free of tapeworms.

Prevention and Monitoring of Fleas

Flea prevention and control in Hawaiʻi is difficult because the warm climate allows fleas to multiply year-round. Flea eggs and larvae are continuing sources of fleas on pets and must be dealt with when attempting eradication. Flea bites on humans may continue months after the host animal has died or left the premises.

If your dog or cat goes outside at all, even just for walks or for bathroom functions, it is likely to pick up fleas. You can't have a flea-free pet for long unless it never steps outside and your entire house is flea-free. To fight fleas, especially in Hawaiʻi, you need a household flea management program. This involves monitoring and preventive techniques to keep flea populations low and a plan of attack for when fleas become unmanageable. The following monitoring and prevention methods will help minimize or eliminate the need for chemical controls.

Flea Traps

Flea traps are sold in pet stores and pet catalogs. They usually consist of a plastic tray lined with replaceable sticky paper and a light source to attract adult fleas. Sometimes a scent is impregnated into the paper that is said to act as an attractant. Fleas that jump into the trap are caught and held on the sticky paper. You can make a homemade flea trap by placing a desk lamp on the floor with a bowl of soapy water underneath it. The fleas are attracted to the light, jump up, fall in the soapy water, and drown.

These devices can trap large numbers of fleas if the population is high. Their main benefit is as a monitoring device to determine whether there are fleas present and in what numbers. A trap can be moved to various parts of the house to determine flea presence in different areas.

Restricting or Controlling Access

Keep your animals on your (hopefully, flea-free) premises; don't let them stray to areas that may be infested. If you take your pet for walks, keep to sidewalks and roads that are unlikely to harbor fleas; don't let your pet stray into yards and grassy areas that may be frequented by other animals. Keep your dog from sniffing strange dogs. This all sounds overly restrictive and not much fun for your pet, but it's part of the price to pay for a flea-free pet.

Flea larvae and eggs are not on your pet but on the floor and in pet bedding. Have your pet sleep in a regular area or special bedding to concentrate fleas in that area. Pay particular attention to that area when cleaning. Wash pet bedding frequently.

Combing/Handpicking

If you have a short-haired cat or dog, a flea comb—a fine-toothed metal comb—is useful for grooming. It is designed to allow the hairs to pass through the tines but catch adult fleas and feces (black specks). It is a useful device for keeping the adult flea population low on your pet. Comb in the direction of the hair, starting at the skin line, and flick any live fleas into a dish or pan of soapy water.

With a long-haired cat or dog, the comb is less effective because the hair gets tangled in the comb. In this case, handpicking is more practical. Part the hair with a comb at the skin line and search systematically a half inch or so at a time. Pay particular attention to the skin at the base of hairs, looking behind the ears, on the feet and between the pads and claws, behind the legs and on the haunches, and on the belly. Pick off with your fingers or comb any flea you see and drop it into a dish of soapy water. This is tedious and takes time, but not only will it keep your animal more comfortable, it will also give you a good idea of the level of your flea problem. Wash your hands after you have de-fleaed your pet as a precaution against accidental tapeworm infestation.

This kind of monitoring followed by application of stronger control techniques will keep flea populations to more tolerable levels throughout the year.

Bathing/Dipping

If your dog has fleas, it should be bathed frequently. This helps kill the fleas and rid the body of flea feces and debris. A mild soap and water is sufficient, especially if you plan to follow the bath later by a chemical dip. You don't need to use an insecticidal soap for bathing; the fleas will drown, especially if the pet is partially immersed in soapy water. Using an insecticidal soap, however, will kill more fleas. See also d-Limolene products below.

Dips really belong in the chemical control section below, but because they most often follow baths (wetting down the animal's hair with a bath before a dip allows the chemical to penetrate to the skin better, especially in long-haired pets), we include them here. Dips are available with mild to strong insecticides. Dips are not usually used on cats because cats are more sensitive to chemicals and because they lick up so much hair in grooming themselves that they can swallow a toxic dose. For dogs, the insecticide (dip) is mixed with water, poured or patted over the pet, and allowed to air dry. The residual continues to kill fleas over several days. If you are not sure about how to use dips or which to use, consult your vet. If you are using dips, it is essential that your pet has not been exposed to other insecticides in collars, dusts, or sprays over a certain prior period; check the label.

Vacuuming/Carpet Cleaning

Vacuuming must be a regular part of the flea management program. For every one flea you find on your pet, there are probably a hundred loose in the house. When you know you have a flea problem, you must vacuum more often. Vacuum not only floors and carpets, but also fabric-covered furniture that the pet may rest on. Make sure that you vacuum cracks and crevices, and pay close attention to where your pet spends its time. After each vacuuming, seal the vacuum bag and trash it immediately. Alternately, freeze it for a few days or expose it to direct sunlight to kill any fleas inside. Vacuuming will remove adults and eggs and food used by larvae, but it is less effective for removing larvae in carpets because they cling to the fibers. You may need to repeat vacuuming at regular intervals, daily if necessary. If the flea population is high and your pet continues to scratch uncontrollably, you may want to shampoo or steam clean the carpet professionally.

Ultrasonic Collars

These are advertised on TV, in pet catalogs, and elsewhere. The collar generates a high-frequency sound that is undetectable by the pet but supposedly unpleasant to fleas; reputedly they will flee from the sound. Ultrasound can repel fleas, but only if the flea is extremely close to (right on top of) the ultrasonic device. Unless you can figure out how to cover your pet's entire body with collars, the fleas will likely only stay away from the collar area.

Control of Fleas

When a flea infestation is firmly established, treating the household pet alone will not control the problem. Treatment must include the yard and rooms in the house that the pet frequents. Treatment must also be repeated at intervals to ensure that all life stages of the flea have been killed. Professional exterminators may have to be called in difficult cases. The Vector Control Branch of the Hawai'i State Department of Health can also help with questions about control.

Fortunately, there are many things a homeowner can try before deciding to call in an exterminator, which can be expensive and, alas, is usually temporary. Some of the following might work in your personal flea management program.

Noninsecticidal Remedies

Yeast. Vitamin B^1 is sold in pet shops and catalogs as brewer's yeast. It has been touted as a flea repellent for many years. It has never worked with our dogs, but some people swear it reduces flea populations on animals. If you want to try it, do so in small doses to be sure it does not cause intestinal problems in your pet.

d-Limolene. d-Limolene flea products are found in dips and shampoos and kill all stages of fleas on contact. This is an extract of citrus peel and is found widely in cosmetics, soaps, perfume, food, and beverages. It is reasonably safe, but may irritate eyes and skin and has caused minor tumors in cats. Apply cautiously at first to a small area of the animal and observe any effects before applying to the entire body.

Once-a-Month Flea Treatment

This book was in press when a new weapon against fleas came on the market. If the treatment, called "Program," works as claimed, it represents a revolutionary breakthrough in flea control.

Program is also available for cats as a liquid suspension. The liquid is available in two sizes: a small pack for cats up to 10 pounds and a large pack for cats 11 to 20 pounds. The liquid is given to each cat in the household once-a-month in food.

Program is manufactured by Ciba-Geigy Corporation and is indicated for use in dogs six weeks or older. Program tablets contain lufenuron, an insect growth regulator. Program does not kill adult fleas but breaks the flea's life cycle by inhibiting egg development. Tablets come in four sizes, meant to be taken orally by dogs and puppies according to their weight. One pill is given once each month with food. The chemical remains in the dog's bloodstream for 30–31 days, at which time another pill must be given. If more than one dog shares a household, all must be treated.

Program can be used along with adult flea sprays and premise treatments. The manufacturer states that the pill alone will bring fleas under control in a few weeks in a mild infestation, in 30–60 days in a heavy infestation. Program is available only through veterinarians and in 1995 cost about $5 per month for each dog, slightly more for each cat.

Dusts

Many dusts are specially formulated for use on pets and are relatively safe if used as directed and with caution. Active ingredients range from safe and environmentally kind (pyrethrum) to stronger and more residual (carbaryl). See the discussion of Dusts and Powders (p. 11).

Some flea-control companies use diatomaceous earth to eliminate fleas. Sometimes this is very effective in controlling fleas. Diatomaceous earth is the bodies of tiny marine animals (diatoms) that are collected as a powder. The dust is applied to carpets and flooring and allowed to sit for a time before the excess is vacuumed up, leaving a residue in cracks and in rugs. The tiny spines on the diatoms pierce the flea's skin, causing it to dry out and eventually die.

The plant product pyrethrum, from the crushed flowers of *Chrysanthemum* species, can be lightly dusted on your pet. It is effective and safe. Pyrethrins, which are the active insecticidal component in pyrethrum, are found in a number of dusts; synthetic versions of pyrethrins, namely pyrenoids, are more toxic. Take care when powdering animals to avoid getting powder in nose, eyes, and mouth of your pet and avoid inhaling the dust yourself.

Silica aerogels can also be used with a fine bulb duster to apply to pets and are also available in combination with pyrethrins for faster kill. Silica aerogels may be difficult to find in Hawai'i, but you may be able to locate Drione, a formulation with pyrethrins. On a very long-haired dog (like a Samoyed, on which we tried Drione), about 2 ounces of powder is needed to cover the entire dog adequately down to the skin. The goal is to be sure fleas contact the powder. Because of the problems with getting adequate coverage on our dog, we were not able to kill all fleas with two applications several days apart. On a short-haired dog, however, it might be possible to cover the animal thoroughly and eliminate all fleas.

We have found considerable resistance in Hawai'i cat fleas to the stronger carbaryl insecticides. Start with the safer dusts first and move on to stronger methods if those don't work.

Sprays

Many sprays available today have synthetic pyrethrins or insect growth regulators (IGRs) as active ingredients. IGRs interfere with the flea's physiology and prevent development of adult fleas from eggs, larvae, or pupae and do so for months after application; adults are not killed. The intent of this product is to break the cycle and prevent the next generation by not allowing the fleas to develop into adults and mate and lay eggs. Products with methoprene, such as Precor, are available readily in Hawai'i and are safe for humans and pets. Sometimes IGRs are combined with conventional insecticides so that all flea stages will be killed; in these cases, the product will not be as safe.

Fleas can develop resistance to IGRs just as to conventional insecticides, so do not overapply them. Another disadvantage is that the adults are not killed and remain around to continue biting and irritating pets and people.

Sprays containing repellents such as pennyroyal, cedarwood, citronella, and others, or pyrethrins may be effective for spraying on your dog's feet, legs, and underbelly before taking it outside for a walk to help avoid transporting fleas into your house from the neighborhood.

Flea Collars

Flea collars are commonly used by pet owners as a convenient form of flea control. The flea collar is a plastic strip impregnated with a poison, usually an organophosphate or a carbamate insecticide. Collars kill fleas on contact or by producing toxic vapors.

Flea collars cause skin irritation in some pets. Moreover, they are frequently ineffective, either because they kill fleas only on the front end of the animal, where vapors are more concentrated, or because fleas have developed resistance. Use caution when handling flea collars, and don't put them on your animal unless it has fleas. When using flea collars, be aware that your pet is constantly breathing poisonous vapors and so are you when you are around your pet.

Foggers/Yard Treatments

IGRs are also found in sprays and foggers used for surface treatments and for carpets. Chemicals intended for treatment of surface areas such as kennels and yards are stronger and should never be put on your pet. Follow label instructions carefully.

If you have a very large flea population spread throughout the house and yard, call in professionals who are licensed for application of the more hazardous insecticides and will be more likely to do a thorough job. After your house and yard have been treated, you can institute a flea management program to keep flea levels low.

FLIES AND MOSQUITOES
Order Diptera

God made the Fly and
Forgot to tell us why
—OGDEN NASH

1,426 fly species in Hawai'i, 1,058 native

Flies are a very large group of insects that range from species as slender and delicate as mosquitoes and crane flies to stout-bodied species like house flies and flesh flies. Flies and wasps look somewhat alike, except that flies have only two wings. Wasps may seem to have only two wings when not flying, because the second pair is often hidden underneath the first pair.

Flies with biting mouthparts, such as the mosquito, pierce the skin to feed on blood. Flies with sponging mouthparts, like the house fly, lap up partially liquefied food. The medical importance of flies and mosquitoes in Hawai'i was covered in *What Bit Me?*

Filth Flies

Families Muscidae and Calliphoridae

Flies such as the house fly, dog dung fly (Muscidae), and green bottle fly (Fig. 61) (Calliphoridae, or blow flies) are called filth flies because the main food source for the larval stage is garbage (waste food), decaying materials, and feces (human and animal waste). Filth flies carry a wide variety of disease-causing organisms and are responsible for many diseases of humans and other animals. These flies do not bite, however.

Flies pass through four stages: egg, larva (often called maggot), pupa, and adult. Adult flies are attracted to garbage and feces to lay eggs. After hatching, larvae feed and grow in these food sources. After larvae reach full size, they crawl out of the garbage or dung and look for a dry place to pupate; the pupa then develops into the adult stage. Depending on the species, this can take 6–30 days.

The number of filth flies you have in your neighborhood is related to the number of breeding and feeding sites. These may be refuse and garbage dumps, but they are mostly the human and animal foods we throw away or

Figure 61. The green bottle fly is in the blow fly family and is important in helping decompose dead organisms. (Photo courtesy of Van Waters & Rogers Inc.)

allow to spoil, as well as other organic wastes, usually pet feces. The key to controlling fly problems is properly managing organic waste materials that flies use for breeding and feeding.

Hawai'i's warm weather speeds up the life cycle of filth flies; the additional warmth produced inside garbage containers, dung piles, and organic composted materials speeds up growth even more and may result in thousands of flies in very short periods.

Filth flies make potentially dangerous housemates. After visiting manure, sewage, garbage, and dead animals and picking up filth on their bodies, they visit your kitchen, dining rooms, and bedrooms, landing on food and drink, and on the lips, eyes, pacifiers, and bottles of children. Don't let these germ carriers share your home with you!

House Fly
Musca domestica
(Fig. 62)

The house fly apparently arrived in the Hawaiian Islands along with the Polynesians, because it was here before contact with Westerners. Although the Hawaiians popped lice and probably fleas into their mouths while grooming each other, flies were utterly detested and avoided if the insects were drowned in their food. Perhaps this indicates that the Hawaiians were aware of the disease-carrying ability of flies.

Figure 62. The house fly multiplies rapidly. It is also a well-known carrier of disease organisms. (Photo courtesy of Van Waters & Rogers Inc.)

Found on all major islands in Hawai'i, *Musca domestica* is closely associated with humans. The fly is gray and about ¼" long. Its back has four black stripes running lengthwise, and the abdomen is marked with white or yellow patches in the male. This is one of the fastest breeding of all insects; a female can lay a total of 600–900 eggs over a period of 4–12 days. Egg to adult takes 6–20 days. An adult may live 1–2 months.

The house fly feeds on anything with a moist surface, including food of all kinds, excrement, and garbage. In nature, these flies are important in breaking down organic debris, such as garbage and manure. Unfortunately, disease organisms are picked up and spread when the fly feeds or lands on contaminated material. Food is then contaminated when the fly walks over it, regurgitates on it, or drops fecal matter.

Dog Dung Fly
Musca sorbens

The dog dung fly is widely distributed through the Pacific and was first found in Hawai'i in 1949. It is now found on all major islands and in many situations is a worse pest than its relative the house fly. The two species look alike. The dog dung fly is slightly smaller, has two dark stripes on its back and gray markings on its belly. The life cycles are similar.

The dog dung fly is important in Hawai'i because the larvae feed on, and break down, dung of all sorts. However, as adults, the flies feed on human food in addition to garbage and filth, thereby transmitting microorganisms. *M. sorbens* is a known transmitter of viruses, bacteria, and parasites. This fly is also attracted to open sores and wounds and may cause infection; it is also attracted to eyes and may transmit certain eye diseases.

Besides its disease-transmitting potential, this fly can be a serious pest inside and outside, in rural and urban areas. It is aggressive and lands on people and animals. Its attraction to sweat, mucous membranes, sores, and wounds can make it extremely unpleasant to be surrounded by large numbers of flies. Its favorite breeding ground is dung—dog, cat, cattle, horse, goat, and pig dung are good breeding sources, with dog dung being the primary breeding material in urban areas.

Control of Filth Flies

Physical Control

Install and keep screens and screen doors in good repair. Seal any cracks or holes through which flies may enter the house. Use a good, old-fashioned fly swatter, a rolled-up newspaper, or even a rubber slipper to kill flies that get into the house. If large numbers of flies are present in the house, you may want to try sticky fly paper, which can be hung in areas where the grossness of seeing fly corpses stuck to the paper is not a bother. There are also other fly traps in the stores, including jars with commercial baits.

Sanitation. Most control measures against filth flies concentrate on strategies for keeping organic wastes properly stored or disposed of so that flies will not be attracted into the area. The basic idea is to keep flies away from wastes by containerizing, bagging, or sealing or to make wastes useless by drying them out to the point that there is not enough moisture to support larval development.

Dog Feces. The dog dung fly does not always stay around its emergence and breeding site; it can move quickly out of its breeding area into the community. One thoughtless pet owner who does not pick up his or her dog's feces can cause fly problems throughout the community. Every homeowner, dog breeder, kennel owner, and dog walker *must* remember to pick up fecal materials left by their animals. The entire community must cooperate in the effort to control the spread of flies.

Stray dogs and cats make the fly situation more complicated, though cats are less of a problem because they tend to bury their feces immediately.

Feces Disposal. Several options are available for picking up and disposing of feces. You can scoop up feces with a tool made expressly for this purpose; this "pooper scooper" is available in pet stores. The feces can then be flushed down the toilet if you are close to home. Otherwise, you can wrap the feces in newspaper to soak up the liquid, put this in a paper bag, seal it, and put the bag in a garbage can or dumpster. If you are not against the environmental consequences of using plastic, put the feces in a plastic bag, seal, and dispose of it.

Nowadays, many people carry plastic bags when walking their dog, invert the bag over their hand, pick up the feces with the plastic-covered hand, strip the bag off the hand while turning the bag right side out, seal, and drop in the nearest dumpster. This is easy and convenient. We have also seen many responsible dog walkers wait for the dog to assume the position, quickly put a newspaper under the dog, then wrap the deposits in the newspaper, and dispose of them.

At home in your yard, you can bury feces immediately after they are laid (if not buried immediately and flies have already laid eggs, burying is not practical because you have to bury the feces very deep [3 ft. or more] to prevent larvae from climbing up through the soil to the surface). You can also compost the feces after drying them out sufficiently in newspaper, sawdust, or sand.

Managing Organic Wastes for Fly Control

✔ Pick up, dispose of, or bury pet feces.

✔ Separate organic from nonorganic, recyclable wastes (glass, cans, paper).

✔ Recyclables should be rinsed thoroughly of all food and liquid before storing or sending out for recycling. Yeast in leftover beer and sugar in cola drinks can attract flies to the area.

✔ Food and garbage stored in the kitchen or pantry should be in sealed containers. This is especially important for fresh foods, leftovers, meats, and the like, whose odors will attract flies.

✔ Waste meant for the garbage can or dumpster should be drained as much as possible; keep in tightly sealed plastic bags until taken to the garbage can or dumpster. Do not leave the plastic bags where strays (or pets) can break them open. For people concerned about the use of nonbiodegradable plastic bags, the garbage can be securely wrapped in absorbent material like newspaper until it is fairly dry; then the wrapped material can be put in a paper bag and thrown away.

✔ Garbage cans should have a tight-fitting lid that will not come off if knocked over by dogs or cats or be blown off by wind. After garbage pickup, wet cans should be hosed out and a disinfectant occasionally added to reduce fly-attracting odors. Don't overfill garbage cans; the lid should fit snugly.

✔ Dumpsters ideally should be kept closed to keep out flies. Garbage and other waste dumped in a dumpster should be wrapped or bagged as described above; it should never be dumped loose into the containers. Dumpsters should also be hosed out regularly and disinfected.

✔ If you prefer to compost your organic wastes, be sure you are informed about composting methods that keep down fly numbers. Periodically turning the compost heap helps.

Chemical Control

A number of insecticides can be used for aerial spraying against filth flies if you have a major fly problem in the house. However, we do not recommend that you use chemical control measures routinely. This only pollutes the air inside your home and will prove only temporarily effective at best.

Other Flies

Vinegar Fly
Drosophila spp.
Family Drosophilidae
(Fig. 63)

These small, yellowish flies (about 1/8") usually have red eyes. They breed in and are attracted to overripe fruits and other fermenting foods. They are often seen hovering in the kitchen, especially when fruits are sitting out. The fly is attracted to the fruit for laying its eggs; larvae feed within the fruit. Yeasts are its preferred food; when they are not available, *Drosophila* will feed on fungi and bacteria. Egg to adult takes 7–30 days; adults can live 13–120 days. This fly does not bite or sting and is mostly a nuisance.

Figure 63. *Drosophila* are known as fruit flies or vinegar flies. The young of these red-eyed flies feed in overripe or rotting fruits. (Photo by Gordon Nishida)

Removal of decaying fruit or refrigerating or containerizing fruit will remove the problem. Also, remove other possible sources of food for fly larvae, such as opened bottles of fruit juice or bottles and cans of beer.

Remember: "Though time flies like an arrow, fruit flies like an apple." (Anon.)

Moth, or Drain, Fly
Psychoda spp.
Family Psychodidae
(Fig. 64)

These short, broad-bodied flies are so woolly or hairy that they look more like little moths (1/16–3/16" long). They are often seen in sinks and on walls and mirrors in bathrooms, laundry rooms, and basements, and around drains, usually where there is sewage. Larvae live in water or moist organic material, feeding on microorganisms, algae, and filth. Moth flies are not harmful; they will not bite. Their presence may indicate a sewage problem or high moisture level in the area. Periodically clean your drains by using a drain plunger or drain clearing fluid to control these flies if they become too numerous.

Figure 64. Moth flies are the original drain cleaners, living on the sludge in your drains. They are really quite elegant looking if you see them close up. (Photo by Gordon Nishida)

Cheese Skipper
Piophila casei
Family Piophilidae

This small (1/6" long), shiny fly, with yellow feet and face, is apparently not established in Hawai'i. However, it is sometimes seen in imported cheese or cured meats, and so we include it here. The young, larval stage is found in cheese or cured or smoked meat. Maggots are small and yellowish and can "jump" by bending their bodies double and snapping them straight. Infested portions of food should be discarded; the remaining, uninfested portions can still be eaten. Meats and cheese refrigerated below 42°F usually will not be infested.

Mosquitoes

That is your trick, your bit of filthy magic:
Invisibility, and the anaesthetic power
To deaden my attention in your direction.
— D. H. LAWRENCE, The Mosquito

Family Culicidae
8 mosquito species in Hawai'i, 0 native

Mosquitoes are well known for their itching bites and for carrying diseases. Only females bite and suck blood; the protein helps mature their eggs. Both females and males feed on plant nectar. Not all female mosquitoes suck blood; *Toxorhynchites* species don't bite, and their larvae prey on the young of other mosquitoes.

Mosquitoes were unknown in Hawai'i until about 1826, when the first species (*Culex quinquefasciatus*) was apparently brought here by a ship from Mexico; mosquito larvae were probably carried inside water casks.

Mosquitoes develop through four stages: egg, larva (wriggler), pupa (tumbler), and adult. Eggs are laid in quiet pools of water; they hatch into larvae in 2–6 days. Larvae feed on microorganisms and pieces of organic matter. Most breathe air directly through a tube at the end of the abdomen; when they need air, they wriggle to the water's surface and hang upside down with the air tube at the surface. Larvae develop into pupae in 5–10 days, and adults emerge in another 1–6 days.

In Hawai'i we have problems with mosquitoes year round because of the warm weather and moist environment. Even with proper screening of the house, because of their small size, speed, and sneakiness adults are able to enter houses quickly as people open doors.

Mosquito Species in Hawai'i

Eight species of mosquitoes are now found in Hawai'i. Five of these are harmful and three are beneficial. *What Bit Me?* dealt in greater depth with the mosquito species in Hawai'i; refer to that book for more information.

Two of Hawai'i's most common species readily enter houses. The Asian tiger mosquito, *Aedes albopictus* (Fig. 65), is basically black, marked with silvery white or yellowish white bands and stripes. The legs are banded with white. This mosquito bites during the day. Its small size and hit-and-run bit-

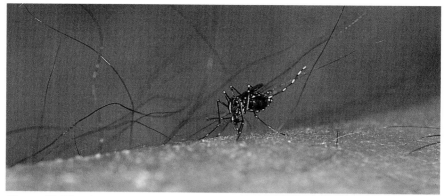

Figure 65. The Asian tiger mosquito is strikingly marked. It can become a major nuisance if not controlled. These mosquitoes are very good at bite-and-run and are active during daylight hours. (Photo by Gordon Nishida)

ing tactics make it a very aggressive and obnoxious pest.

The southern house mosquito, *Culex quinquefasciatus* (Fig. 66), is medium-sized, dull brownish or yellowish in background color, with uniform white bands at the bases of the abdominal segments. It bites at night and ranges in elevation from sea level to 6,000 feet. It can fly up to 3 miles seeking a meal. It is found on all the Islands and is the most common species in Hawai'i. This mosquito commonly breeds in standing water such as ground pools or swamps, but can also breed in artificial containers in backyards.

Figure 66. If your home is not well screened, the southern house mosquito will come in and bite you while you are sleeping. (Photo courtesy of Van Waters & Rogers Inc.)

Detection and Monitoring of Mosquitoes

One of the best ways to keep down mosquito populations is to eliminate their breeding places. Survey your yard for standing water in which mosquitoes might be developing. Check for mosquito larvae; you should be able to see the larvae wriggling in the water. Most of our mosquitoes can breed in artificial containers like cans, bottles, old tires, and anything that collects water. Check also the following: birdbaths, clogged gutters, puddles formed from air conditioners, dripping or leaking faucets, ponds, tree holes and stumps, holes in rocks, and wheelbarrows. Another good breeding place is the drip pans kept under plants, especially if they fill with standing water. Keep these containers dry or put a few drops of light machine oil or mineral oil onto the standing water (see Oil Films under Chemical Control, below). The oil prevents the mosquito larvae from penetrating the surface of the water to breathe. Plants like bromeliads and pitcher plants that hold water also attract egg-laying females; mineral oil also works in these. Unfortunately, in many parts of Hawai'i, rain will fill the container and cause the oil to run off.

Sources of mosquitoes can be inside the house if there is standing water where females can lay their eggs. This might be unused toilets, plant drip pans, and containers for rooting pineapple or avocado or other plants.

The best method of controlling mosquitoes is to eliminate all the sources of standing water. Drain, fill in, discard, cut up, turn upside down—whatever will work to eliminate the standing water in containers. Even small amounts of water, like that at the bottom of a dog dish, can be a source of lots of mosquitoes. Where you cannot drain the water, see other measures below.

Control of Mosquitoes

Physical Control

Take the obvious precautions. Screen windows and doors and keep them in good repair, fixing any holes in the screen or gaps around screens and window frames. In situations where screening is impractical, fine netting over your bed can be used to protect you while sleeping. Also, head nets can be worn in heavily infested areas to protect the face and neck from bites.

Biological Control

Mosquito-Eating Fish. Backyard ponds that support mosquito breeding can be stocked with goldfish or mosquito-eating fish like guppies or *Tilapia*. Call your vector control agency to ask what is recommended for your situation and area. They may also keep stocks of certain fish to help you with mosquito problems. Please be sure not to release these fish into our streams because they become problems, affecting our native species of fish and insects.

Bacillus Thuringiensis Israelensis (BTI). BTI is a commercially available bacterial species that kills several kinds of insects. It was first used in commercial formulation against moths in the late 1950s and has since been developed for use against larvae of mosquitoes and other flies. BTI is considered harmless to humans and other animals, as well as to natural enemies of target species.

BTI comes in powder or granular form, is added to the water where mosquitoes develop, and acts as a stomach poison when larvae eat it. It kills larvae rapidly (in 15 minutes when applied in good doses) and breaks down rapidly in the environment.

Unfortunately, BTI is not available in Hawai'i in amounts economically feasible for use by the homeowner. In any event, it is most effective when applied by vector control agencies on a community-wide basis.

Chemical Control

Oil Films. Applying an oil to the surface of water so that it forms a film will kill any mosquito larvae present. The oil clogs their breathing tubes and suffocates them, because they cannot penetrate the oil to breathe. Some vector control agencies recommend using a light machine oil, but heavier oil can also be used. Beware that most oils will pollute the water and may kill nontarget organisms like fish and plants. Special oils specifically for mosquito control are now available, which are less polluting and damaging to nontarget wildlife.

Repellents. Repellents keep mosquitoes away from your body. It makes good sense to protect yourself against mosquito attack, especially outdoors, rather than pollute the air with sprays, which are temporary and ineffective.

Deet (diethyltoluamide) is the most widely used and effective repellent for a variety of insects, including mosquitoes (see discussion on page 15). Unfortunately, deet is absorbed into the bloodstream and can have serious medical consequences. Rashes are common side effects. Apply a test patch a few days ahead of time to see if you are sensitive before giving yourself the full treatment. Deet preparations on the market range from 20% to 95% deet content. It is best to use a low-deet repellent and put on more if insects continue to attack. Very high content deet preparations are very effective against mosquitoes for many hours, but they contain more deet than needed. On children, it is best to use repellents with no more than 20% deet.

If you are sensitive to deet, you may wish to treat clothing instead of skin. Check what fabrics the deet preparation is safe to use on. Cotton, wool, and nylon should pose no problems.

Smoke coils containing pyrethrins repel mosquitoes and are used commonly in Hawai'i. They are effective and safe to use outdoors. Use them very cautiously indoors, and only in a well-ventilated area.

Citronella candles containing citronella oil can be used outdoors for some protection in small areas. They are not very effective in large areas under attack by aggressive mosquitoes.

Sprays. These are widely available in pyrethrin formulations for use as space sprays inside and outside the house. Sprays are useful in emergencies inside the house as temporary stopgap measures. But unless you solve the problem of mosquito breeding in your area and entrance into your home, spraying will be only temporarily effective.

Why Do Mosquito Bites Itch? Why Do Mosquitoes Bite Me and Not My Husband?

Mosquitoes inject saliva as they bite. The saliva acts as an anticoagulant so that blood will flow easily. The saliva causes swelling and itching that varies in intensity from person to person. Some people may develop some immunity and react hardly at all, but others may have swelling beyond the bitten area. Anti-itch remedies are available in drugstores. If the itching is serious, consult a doctor or skin specialist.

Why do mosquitoes seem to bite some people and ignore others? Several factors are at work here. There is no question that some people are more attractive to mosquitoes (and fleas) and are bitten more frequently. This has to do with skin secretions, such as sweat and lactic acid, and emanations, such as carbon dioxide, which we exhale. Also, our individual immune systems vary and so does sensitivity to an insect's saliva. Some people are allergic to bites from mosquitoes and fleas and suffer more discomfort, even though they may not be getting bitten more often than people around them.

GRASSHOPPERS, KATYDIDS, AND CRICKETS
Order Orthoptera

These are not usually thought of as household pests, but occasionally they wander into houses. Sometimes they annoy people with their sounds and are the source of questions and complaints by householders in Hawai'i.

Grasshoppers, crickets, and katydids usually have the ability to "sing" or at least make some sort of sound. The grasshoppers are the poorest singers; some are barely able to make scratching noises. Some grasshoppers make sounds by "crackling" the wings while flying. Crickets are well known for making noise at night; in some parts of the world, especially in the Orient, they are kept as pets because their singing is believed to bring good luck. Cricket songs are usually the most musical. Katydids often make the loudest and sometimes the harshest sounds.

The insects we know as katydids, or long-horned grasshoppers, are sometimes called grasshoppers. Technically, grasshoppers (short-horned grasshoppers) actually have short antennae and the hearing organ or "ear" on the first segment of the abdomen. Katydids (long-horned grasshoppers) have antennae that are usually longer than the body and have hearing organs located on the forelegs.

Grasshoppers
Family Acrididae
5 species in Hawai'i, 0 native

People occasionally are curious about the large short-horned grasshopper we have in Hawai'i. It gets to be 3–4" long and is called the vagrant grasshopper *(Schistocerca nitens)* (Fig. 67). Though related to desert locusts, which were the scourge of biblical times and are still a problem in Africa, the vagrant grasshopper does not seem to do much in Hawai'i except for occasionally harming plants in gardens. It rarely, if ever, enters homes.

Figure 67. The vagrant grasshopper is a large, brownish gray grasshopper that occasionally becomes a pest in gardens. (Photo by Gordon Nishida)

Figure 68. The aggravating grasshopper makes a long, shrill buzzing noise. (Photo by Gordon Nishida)

Katydids
Family Tettigoniidae
20 species in Hawai'i, 12 native

One katydid found in Hawai'i is known as the aggravating grasshopper (even though it is a katydid!) because its song is loud, shrill, and continuous, sounding somewhat like a high wind whistling through telephone wires.

The aggravating grasshopper *(Euconocephalus nasutus)* (Fig. 68) belongs to the group known as coneheaded grasshoppers, which are named because the top of the head, between the eyes, is usually cone-shaped or somehow expanded. The loud, shrill, continuous calls of this long-horned grasshopper irritate some people. They are also excellent ventriloquists, as they seem to "throw their voices" when you are out in the bushes looking for them.

You will know these critters are around by the long, steady buzzing sound. They are usually found in brushy areas with low vegetation and tall grass and may sometimes feed on plants in gardens or yards.

Crickets
Family Gryllidae
252 species in Hawai'i, 243 native

Crickets are related to grasshoppers and katydids. You can tell them apart by the way the wings are held over the back. In grasshoppers and katydids, the wings are held "rooflike," with the top edges of the wings touching or slightly overlapping and sloping downward. Crickets have their wings flat and overlapping on their backs. Female crickets also have an egg-laying device (ovipositor) that is spear-shaped or arrow-shaped; the katydids have ovipositors that are sword-shaped.

Crickets are active at night, foraging for food and looking for mates. Crickets feed mostly on vegetable matter, though if given the opportunity will feed on many other things including insects. They have been reported to chew on clothes and, if given the chance, will attack foodstuffs.

Though there are many native species of crickets, they are mostly found in forests and will not affect you. The species listed below may occasionally become house pests.

The **house cricket** *(Acheta domesticus)* (Fig. 69) is the only cricket known to be directly associated with houses. It is about ¾" long and light yellowish brown, with three dark bands across its head. Cricket eggs are whitish yellow, tiny, and laid in cracks and crevices or in soft soil. Egg to adult may take 7–30 weeks. These crickets are fond of warmth, so in mainland areas, they often come inside to get out of the cold. In Hawai'i, they do not seem to be abundant and pretty much stay outdoors; you might find them around your water heater or under a refrigerator. They like hiding in cracks and crevices, but most modern buildings do not offer them many hiding places. The house cricket is usually a scavenger, but can become a pest in the house by eating unprotected food. They prefer softer foods, but will nibble on many items, damaging them in the tasting process. They can bite.

Figure 69. The house cricket is yellowish brown and is the most common cricket found in homes. Here, adults and nymphs feed on a wool blanket. (Photo courtesy of Van Waters & Rogers Inc.)

Figure 70. The field cricket is black or blackish brown and is usually found outdoors under objects (stones, cardboard boxes on the ground, and other such things). Shown is the male *(left)* and female *(right)*. (Photo courtesy of Van Waters & Rogers Inc.)

The **flightless field cricket** *(Gryllodes sigillatus)* is mottled brown and $3/8-1/2$" long. Neither the males nor females fly; their wings are reduced to small lobes. The male's wings are larger (but only cover about half the body) and are used in singing.

The **oceanic field cricket** *(Teleogryllus oceanicus)* (Fig. 70) is larger, $3/4-1$" long, yellowish brown to blackish, with a darker brown head and thorax. These crickets rarely enter the house, but are more likely to be outside, occasionally on the lānai or at the base of the house, under stones, boards, trash, or in cracks, or under the house if it is elevated. They feed on organic material and sing at night.

MILLIPEDES
Subclass Diplopoda

26 species in Hawai'i, 16 native

Millipedes differ from centipedes in having two pairs of short legs on each body segment, a tubular or cylindrical body, no poison fangs, and a vegetarian feeding behavior. They often curl into a ball as a defensive move. Millipedes do not bite or sting, but can emit a spray or droplets of toxic fluid that may burn the skin and cause injury to the eyes of humans and animals. The fluid is used as a defense against predators.

Rusty Millipede
Trigoniulus lumbricinus
(Fig. 71)

This millipede is about 2" long and has more than 90 pairs of legs. It is reddish brown as an adult, pale whitish when young. It is rarely found in the house. It frequents composts and trash piles outdoors, where it feeds on organic materials. If found inside the house, the poor millipede is usually lost; escort it outside rather than killing it.

Figure 71. Millipedes like the rusty millipede usually protect themselves by coiling into a ball, with their head in the center of the coil. (Photo by Gordon Nishida)

No common name
Spirobolellus sp.
(Fig. 72)

Description: The actual name of this species is not clear and research is underway to determine the true identity of this millipede. *Spirobolellus* ranges in color from cream to reddish with two light reddish or pinkish bands surrounding a thinner dark band running the length of the top of the tubular body. *Spirobolellus* may be up to 1½" in length.

Evidence: Physical presence.

Figure 72. Millipedes also protect themselves by squirting a chemical. *Spirobolellus* can cause minor chemical burns on exposed skin. (Photo by Gordon Nishida)

Impact: This millipede is occasionally found in homes seeking moisture and food. Millipedes usually feed on organic materials and do not bother humans. However, this species has caused chemical burns on exposed parts of the body when provoked to release its defensive secretions. The effects of the chemical include staining and blistering. In homes, millipedes may stain carpets and flooring reddish purple with their secretions.

Preferred Food: Organic materials.

Preferred Locations: *Spirobolellus* is usually found under lumber, rocks, refuse, or wandering about, usually on the ground outside. It may enter homes. It is active at night and may be found climbing walls.

Prevention: If extremely numerous, try to find where the millipedes are entering the house and either caulk or seal the access.

Control: Physical removal is often all that is necessary. Most people can pick up the millipede with their fingers and come away with just a stain. Other areas of the skin seem to be more sensitive to the spray; eyes are particularly vulnerable. Be cautious when killing or moving these creatures outside.

MITES
Class Arachnida: Subclass Acari

> *The life of* Dermatophagoides *is literally dust to dust;
> they're born to dust, live in dust, and die in dust.*
> — BERENBAUM

521 species in Hawai'i, 123 native

Mites are tiny, eight-legged animals closely related to spiders. They are often only barely visible to the naked eye. Many are important in transmission of diseases. They may bite humans when they infest foodstuffs and furniture or domestic pets. They can cause itching and skin inflammation, and may produce allergic reactions. Mites most commonly causing problems in the home are discussed here, with suggestions for prevention and control.

Rat and Bird Mites

Tropical Fowl Mite
Ornithonyssus bursa
Northern Fowl Mite
Ornithonyssus sylviarum
Tropical Rat Mite
Ornithonyssus bacoti (Fig. 73)

In Hawai'i, several species of mites are known to bite humans. The most common problems in the household involve the fowl mites *Ornithonyssus bursa* and *O. sylviarum* and the rat mite *O. bacoti*. These mites are found throughout the world associated with bird or rat nests and occasionally bite humans when their populations are high or their usual bird and rodent hosts are unavailable.

Description: These mites are very small, less than 1/16" long, light gray or yellow. They may be seen by the naked eye when moving about or when numerous.

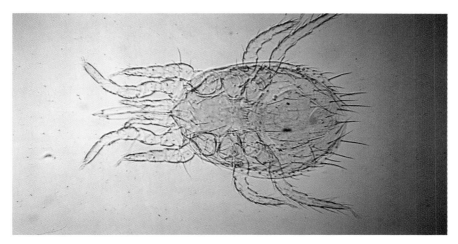

Figure 73. Tropical rat mites are usually found on rats but will bite people if their usual food source disappears. (Photo courtesy of Van Waters & Rogers Inc.)

Evidence: Mite bites (Fig. 74) are often difficult to diagnose and are sometimes mistaken for flea bites. Mites inject saliva when they bite, causing irritation, intense itching, and occasionally inflammation of the skin. The bites appear as small, white blisters surrounded by redness. You may see tiny, colorless creatures crawling around on your skin or you may not see any evidence of what is biting you.

Impact: These mites are not able to sustain themselves on humans and do not transmit any diseases. However, bites may be very itchy, and scratching may lead to infection. In some cases, there may be fever and other symptoms such as headache, nausea, vomiting, loss of appetite, and diarrhea. If symptoms are severe, see a doctor. Topical anesthetics such as hydrocortisone cream, Benadryl, or salves containing benzocaine may help the itching.

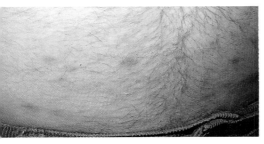

Figure 74. Tropical rat mite bites. (Photo courtesy of Van Waters & Rogers Inc.)

Preferred Food: Fowl mites are parasites of birds and rat mites of rodents; they suck blood.

Preferred Locations: Fowl mites are usually associated with bird nests, and the rat mite is usually found on rats and in their nests. Most fowl-mite problems for humans in Hawai'i occur in the late spring or early summer as birds make their nests in and around homes. When fledglings leave the nest or birds die, fowl mites may invade homes looking for a host. The rat mite frequently attacks humans when rats die as a result of trapping, poisoning, or destruction of their nests.

Prevention/Control: These mites enter houses easily if nests are close to the house, in gutters, or in attics, or if tree limbs touch the house and provide a route for hungry mites looking for a blood meal. Sources of the pest mite, such as bird or rat nests, should be removed, and means of access to the house must be modified or destroyed. The areas infested outside and inside may have to be chemically treated. Destruction of rats or rat and bird nests may provoke additional mite attacks at first. Thus, if bites are severe, several members of the family are affected, or large areas of the house are involved, it is best to call in professional vector-control officers or exterminators to assist with removal of nests and with chemical control.

Stored-Products Mites

A number of mites are minor pests of stored food and of furniture upholstered in certain fibers. When populations are large, these mites can damage grain products such as flour, cereal, corn, dried vegetables and fruits, and cheese. In addition, they may contaminate the foodstuffs with their droppings and dead bodies. Mites may also get on the skin and cause an itchy skin condition (dermatitis) in humans.

The house mite, *Glycyphagus domesticus*, develops in flour, wheat, hay, sugar, cheese, and other products. It can get on the skin and try to feed, but it cannot develop on humans. It is most often associated with people working with flour, meal, sugar, dried fruits, copra, cheese, or ham in factories and grocery stores. These people may develop what is known as grocer's itch.

The straw itch mite, *Pyemotes boylei*, is usually a parasite of insects infesting grain or seeds; it occasionally invades homes. In Hawai'i it is more often found outside infesting seedpods of *koa haole, kiawe,* or monkeypod. The straw itch mite occasionally becomes a problem after treatment of homes for termites. These mites feed on the dead termites, build up large populations, and attack humans when the termite food runs out. Humans are also bitten by this mite when they handle infested seeds or seedpods or occasionally if the mites are blown on to them by the wind. A delayed dermatitis may result from the mite bite, with welts and inflammation.

To guard against stored-food mites, check all dried foods before purchasing; if foods are already in the home, throw away any materials that look suspicious or that have tiny animals moving within them. Keep cereal products dry and store them in sealed containers.

House Dust Mites

By now most of you are familiar with the fact that dust in the home contains many living things; the most notorious of these are house dust mites. These microscopic mites do not bite or sting, but they affect people's health. They are found throughout the world and in almost all households, but are probably more common in Hawai'i and other tropical and semitropical areas.

What Is House Dust?

House dust is a term used for small particles that can become airborne. Common components of house dust include fibers from clothing and paper, ash, fingernail and toenail clippings, food particles, hair, skin scales of humans and pets, paint particles, plant parts, pollen, crystals, fungal spores, soil, and wood shavings. Many insects and other arthropods also live in house dust: booklice, silverfish, sowbugs, human itch mites, cockroaches, beneficial predators, and, of course, house dust mites.

House Dust Mites
Dermatophagoides spp.
(Fig. 75)

Description: These mites are tiny—invisible to the naked eye. Species of the genus *Dermatophagoides* are often the most common mites found in vacuum cleaner sweepings, although other mites, including those that prey on house dust mites, are also found.

Evidence: There are probably few if any households in Hawai'i where you cannot find these mites. There are no visible signs. If, however, you were to take vacuum samples from your bedroom carpets and beds and examine these under a microscope, you would see many living creatures—including house dust mites.

Impact: House dust may trigger asthma and other allergic conditions in

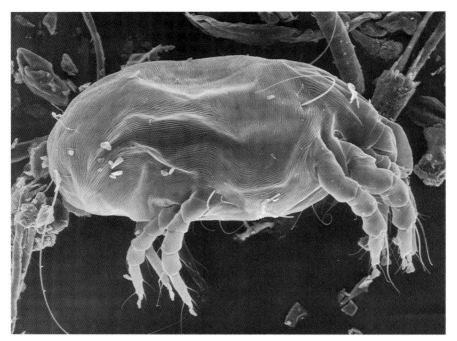

Figure 75. House dust mites are so tiny that they can't be seen with the naked eye. This photograph was taken with the aid of a scanning electron microscope. (Photo courtesy of Dennis Kunkel)

susceptible persons. This includes atopic dermatitis, a skin disorder usually found in infants and children. There also has been research into whether house dust may be involved with sudden infant death syndrome (SIDS). Studies have shown that the mites themselves, parts of mites, their excretions, and their shed skins serve as the allergens in house dust. If you or family members suffer from asthma or similar respiratory problems or have skin conditions that are undiagnosed, consult a dermatologist to determine if house dust mite allergy might be involved.

Preferred Food: These mites feed directly on particles of house dust—pollen, spores, fungi, plant fibers, insect scales, and animal and human dander. They are even known to feed on human semen on bed sheets.

Preferred Locations: House dust mites live in the ever-present layer of minute dust particles that covers everything in houses, including floors, carpets, shelves, furniture, and beds. The favored sites for mites are beds, which have the high-humidity conditions these mites require, as well as a ready source of food in the form of shed skin scales and other human products. When temperature and humidity are favorable, they will develop in other areas of the house, such as in carpets and overstuffed furniture.

Control: House dust mites are everywhere. They are probably exceedingly abundant in Hawaiian households that are not air-conditioned. It is impractical and, indeed, impossible to try to eliminate them from your house. You should make the effort to reduce their populations only if you know that dust mites are causing problems.

There are a number of ways to minimize dust mite populations. These measures will do the most good in the bedroom or sleeping areas. Avoid upholstered furniture. Reduce wall-to-wall carpeting; use washable area rugs. Use curtains and bedding that are washable. Avoid a clutter of objects around the room that will gather dust. Encase mattresses and pillows in plastic covers; there are special products on the market for this purpose. Avoid having furry or feathery pets (dogs, cats, birds, for example), whose hair and dander provide additional food for mites. Keep household, especially bedroom, humidity, low (below 70%) by using dehumidifiers or air-conditioners. Most important, vacuum: thoroughly vacuum mattresses, box springs, pillows, floors, curtains, and other items at least weekly. Wash in hot water or dry clean all bedding twice monthly.

Tests are available to determine whether a person is sensitive to house dust, and, if so, additional preventive or control measures may be recommended. Consult a doctor or allergist if you suspect you are allergic to house dust.

MOTHS
Order Lepidoptera

1,438 species in Hawai'i, 948 native

Moths are stout-bodied, furry, usually plain-colored insects. Moths are often attracted to lights and enter homes if entryways are available. Most moths feed on plants as caterpillars and are not likely to become pests inside the home. However, some small moths are pests of carpets and fabrics, and others are pests of cereals and grains. Case-building species, like the grass bagworm and the household casebearer, are not pests but because their cases are very obvious and the source of many questions, we have included them below.

Grain and Cereal Moths

Stored human foods offer an abundant supply of food for many pests. These provide all the nutrients needed, a friendly environment, and an almost unlimited food supply. Most of the pests of human foodstuffs are beetles, mites, and moths; beetles and mites are discussed elsewhere in this book. See also the discussion of Stored-Food Pests in the beetle chapter.

Most pest moths are tiny, with a wingspan of about ½", and are pale gray, buff, or yellowish brown. Moths fluttering around in the kitchen or pantry are probably stored-products pests; moths flying around in bedrooms or closets are probably clothes moths, discussed later. Many moths, perhaps up to 80 species worldwide, live and feed in a wide variety of stored-food products: grains, nuts, candies, cereals, spices, dried fruits, crackers, pet food, tobacco, drugs, legumes, meats, and so forth.

Moths and beetles are both responsible for major losses of stored food caused by their actual feeding, the presence of their body parts, and spoilage. There are some differences between moth and beetle damage, however. Moth larvae of many species spin dense webs in and over the food. Beetles do not. Some moth larvae tend to leave the food to pupate and may be seen wandering about on walls and ceilings. Beetles spend all of their life cycle in the food product, although adults may escape and look for new food to invade. Adult moths emerge and fly around in kitchens and

Figure 76. The Indianmeal moth is recognizable by the broad gray band that covers its bronze-colored back. (Photo courtesy of Van Waters & Rogers Inc.)

pantries and thus can be swatted or sprayed with insecticides, but always be careful with the use of pesticides around food items.

Indianmeal Moth
Plodia interpunctella
(Fig. 76)

Only relatively few species of stored-products moths are common in households but they can be very abundant and visible. The Indianmeal moth is occasionally introduced into the home in cereal or cereal products and is one of the most common insects in stored products—literally from dehydrated soup to nuts. Details follow as an example of one of the many moth species that infest stored products. Other commonly encountered species are the Angoumois grain moth *(Sitotroga cerealella)* and the Mediterranean flour moth *(Ephestia kuehniella)*.

Description: The Indianmeal moth is a rather elegant-looking moth, with a wingspan of about ⅝", basically pale gray with the last two-thirds of the wing tips metallic coppery or reddish brown. The caterpillars take on the color of the food and may be white, yellow, or even pink or green, with a brown head. Larvae feed within webbing but crawl out to seek a place to pupate. The life cycle may take from 30 to 300 days depending on the kind of food and the temperature; shorter periods are common in Hawai'i.

Evidence: The small (up to ⅝" long) caterpillars (Fig. 77) spin a web over the food object. Webbing and the presence of caterpillars confirms that you have an infestation of moths.

Impact: This species can be a destructive pest by feeding on the food item or soiling the food with its droppings and cast skins.

Preferred Food: This moth feeds on great variety of grains and cereal products and also dried nuts and fruits, chocolate, and candies.

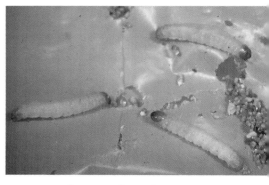

Figure 77. The caterpillars of the Indian-meal moth web together clumps of their food mixed in with their droppings. (Photo courtesy of Van Waters & Rogers Inc.)

Preferred Locations: Caterpillars are found within the food item. Adults may be seen flying about the room during the evening. They usually avoid lights.

Prevention: This pest is not a serious problem in Hawai'i. These moths are introduced occasionally in food products when sanitation or pest control at the original packaging site does not prevent occurrence.

Control: If you catch the infestation early enough, you can try freezing the food item for 15 days in your household freezer. If the food is too far gone to save, seal in a heavy container before discarding, or freeze the food before throwing it away to kill any remaining caterpillars and eggs so that they can't spread. Be sure to check all stored products on your shelves and vacuum up any spilled food particles.

Clothes Moths

Casemaking Clothes Moth
Tinea pellionella (Figs. 78–80)
Webbing Clothes Moth
Tineola bisselliella (Figs. 81–82)
Carpet Moth
Trichophaga tapetzella

Figure 78. The casemaking clothes moth adult is not distinctive looking, but it does have a fluffy hairdo and a black spot in the middle of the wing. The larvae do the damage; adults probably do not feed. (Photo courtesy of Van Waters & Rogers Inc.)

Figure 79. Casemaking clothes moth larvae spin a silken case that they drag around with them. They make new, larger cases as they grow. (Photo courtesy of Van Waters & Rogers Inc.)

Figure 80. Casemaking clothes moths damage materials by chewing small holes in them. (Photo courtesy of Van Waters & Rogers Inc.)

Clothes moths are widely distributed around the world. The casemaking clothes moth and webbing clothes moth are two of the most common species; the carpet moth is less common. Clothes moths are so named because their caterpillars feed on clothing and fabrics, as well as on many other items of animal origin. Only caterpillars do damage; adults do not have a functional gut and do not eat.

Description: White caterpillars (larvae) (Figs. 79 and 82) are tiny, ranging from $1/16$" in early stages to $1/3$" when fully developed. Larvae can take from a month to 4 years to develop depending on the temperature and other conditions. Adult moths are cream or buff, almost golden-colored, small, fragile, with no more than a $1/2$" wingspread. The fluttering flight of the adult moths is distinctive; food-infesting moths have steadier, calmer flight. Clothes moths have a pattern of fly, land, walk... fly, land, walk... and are easy to catch in a jar or vial. Adults live 2–3 weeks.

Evidence: Damaged fabrics have small holes eaten in them by tiny, white caterpillars (Fig. 80), usually accompanied by eggs and silken cases or webbing. Holes are usually scattered about the garment; carpet beetle holes are usually large and concentrated in a few areas. Moth damage is usually cluttered with droppings and cast-off skins, in contrast to carpet beetle damage, which is fairly clean and free of debris. Tiny moths may be running about in damaged areas or flying around the closet or room.

Impact: These are destructive pests that can damage or completely destroy items they feed upon.

Preferred Food: Caterpillars feed on wool, hair, feathers, furs, and other animal products, as well as upholstered furniture, dead insects, and dead animals. Adults do not feed.

Preferred Locations: Adults avoid light and attempt to hide when disturbed. Clothes moths are not attracted to light, but flutter around in dimly lit areas. Fabrics left undisturbed for long periods and stored in dark places are the most likely to be attacked.

Prevention: Caterpillars cannot develop in clean fabrics. They need food stains, beverages, sweat, or urine to get the essential nutrients

Figure 81. The webbing clothes moth looks very much like the casemaking clothes moth except it doesn't have a black spot in the middle of the wing. (Photo courtesy of Van Waters & Rogers Inc.)

Figure 82. The caterpillars of the webbing clothes moth do not spin a silken case but spin a silken feeding tunnel or sometimes just a patch of silk around the feeding area. (Photo courtesy of Van Waters & Rogers Inc.)

to complete development. As a preventive measure, garments should be cleaned before being hung in a closet or stored. Potential moth food items should be stored in tightly sealed containers and aired out occasionally. Moth eggs, larvae, and cocoons are fragile; shaking, brushing, or drying in the sun will remove them from materials or kill them.

Control: Chemical control is not usually necessary, although the use of chemicals as repellents could be effective in protecting your stored clothing. Three chemicals are sold for protection against clothes moths: camphor, naphthalene, and paradichlorobenzene (PDB). Camphor is effective as a repellent and a fumigant and is less poisonous than the other two; it does not build up in fatty tissues. At Hawai'i's normal average temperature and humidity, PDB can act as a fumigant, actually killing the pests. Naphthalene

Figure 83. Clothes moth damage to museum collection. (Photo courtesy of Van Waters & Rogers Inc.)

repels, but does not usually kill. It is important that the container you use to store the fabrics be sealed airtight; otherwise the fumes cannot build up to sufficient levels to be effective. If you choose to use any of these chemicals, air out clothing and clean them before wearing to minimize exposure to these potentially hazardous chemicals (see discussion on p.15).

Cedarwood or cedar oil have been used for many centuries, and many people swear that fabrics stored in cedarwood chests do not seem to get infested. Cedarwood or cedar oil are only effective in a chest with a very tight-fitting lid. The cedar oil vapors will kill newly hatched clothes moth larvae, but will not kill older larvae. The benefits of the cedarwood appear to last only about 3 years.

Moths Occasionally Found around the Home

Black Witch
Ascalapha odorata
(Fig. 84)

The black witch is a very large, black moth, measuring over 4" in wing span. Females are paler in color than the darker male. These moths occasionally enter buildings, garages, and lānais and perch up high in a dark corner of

Figure 84. The black witch moth is sometimes found resting in the house during the day. Color variations found on different islands are shown here. (Photo by Gordon Nishida)

the room, avoiding sunlight during daylight hours. Caterpillars feed preferentially on monkeypod tree leaves at night and hide in cracks and crevices and under bark during the day.

If you find the moth in your house, catch it if you can and release it outside. It is harmless. Control is not necessary unless the caterpillars become serious pests defoliating your monkeypod trees.

Figure 85. Household casebearers are fairly common in homes in Hawai'i. The head of the caterpillar peeks out from the top of the case, which is usually attached to a wall or ceiling. (Photo by Gordon Nishida)

Household Casebearer
Phereoeca allutella
(Fig. 85)

This moth's caterpillar builds a flattened, boat-shaped or almond-shaped case (bulging at the center), with a hole at either end. The cases are about ½" long and often have little bits of debris woven into them. Caterpillars drag their cases across floors and up walls. Cases are frequently seen attached to walls, because caterpillars often climb walls before they emerge as adults. Caterpillars feed within their cases on insect parts and dust or other bits of organic material on the ground until they are fully grown. They prefer darker or damper rooms in the home.

Figure 86. The grass bagworm is a fairly recent immigrant to Hawai'i. Where they are abundant, walls may be covered by these small, odd-looking cases. (Photo by Gordon Nishida)

Grass Bagworm
Brachycyttarus griseus
(Fig. 86)

The larval stage of this moth makes a spiky-looking case about ¼–½" long from pieces of plants, usually grass, and silk. When ready to mate, the female climbs a wall and attaches the case; she does not fly. These little cases may be found attached to house walls. Unless very numerous, they do not seem to cause much damage.

SILVERFISH
Order Thysanura

7 species in Hawai'i, 0 native

These are small, carrot-shaped, silvery creatures that move quickly like wiggling, swimming fish, giving them the name "silverfish." They are also known as bristletails because of the three long strands at the end of the body. These insects love starch and are attracted to books by the paste in the bindings and glue (sizing) in the paper. They don't usually do much damage, but if populations are large enough they can badly damage paper goods, including books, and fabrics.

Urban Silverfish
Ctenolepisma longicaudatum
(Figs. 87, 88)

Description: These insects are silvery gray and the body is carrot-shaped (Fig. 87). They are wingless, the antennae are long and thin, and the tail usually ends in three long strands. They are up to ½" long and may live up to 8 years.

Evidence: Paper products and glue sources such as bindings of books, paper, cards, boxes, starched clothes, and similar materials have the surface scraped off in uneven patches (Fig. 88). Scattered grayish brown fecal pellets may be seen. Silverfish prefer to be on or close to floor level, but will damage wallpaper, especially if a starch-based paste is used to mount the wallpaper; they are occasionally found in cupboards or bookcases. If a box or stack of papers is moved, these silvery gray insects will scurry around to find another hiding place.

Figure 87. Urban silverfish resemble little minnows as they wriggle across the floor. Their silvery color and three long tails help you recognize this occasional pest. (Photo by Gordon Nishida)

Figure 88. Silverfish damage materials by chewing small, irregular holes or scraping off the surface in patches. (Photo courtesy of Van Waters & Rogers Inc.)

Impact: If left long enough, paper will have lots of holes. Glued items may be destroyed by feeding silverfish, and books and starched material may be damaged.

Preferred Food: These insects prefer starchy materials, vegetable matter high in carbohydrates such as flour and oatmeal, and cloth of vegetable origin such as linen, rayon, and cotton. Silverfish rarely feed on cloth of animal origin such as silk or wool. They like glue on the back of stamps and labels.

Preferred Locations: Silverfish like dark, warm, humid places such as closets, drawers, and storerooms. They are more likely to be found where things lie undisturbed for some time and are active at night.

Prevention: Be careful not to bring home bags and boxes that harbor silverfish. Remove potential sources of infestation such as stacks of paper or cardboard boxes on the floor. Mount wallpaper with nonstarch-based paste. Store starched clothes on hangers in closets or in upper drawers of bureaus, armoires, and similar places. Seal up cracks and other hiding places.

Control: Chemical control is usually effective but usually is not necessary; read the label on the insecticide before applying. Boric acid powder and other dusts are effective, as is diatomaceous earth.

Silverfish
Lepisma saccharina

This species looks very similar to the urban silverfish, but is smaller, not quite ½" in length; it is mottled with patches of whitish and blackish marks. It may live up to 3 years. The silverfish is not as important a pest as the urban silverfish and prefers protein such as dried beef.

SPIDERS
Class Arachnida: Subclass Araneae

"Will you walk into my parlour?"
said the Spider to the Fly.
— HOWITT

197 species in Hawai'i, 118 native

Spiders are creatures we both fear and hate. It's too bad; of the thousands of species in the United States (and the couple of hundred species in Hawai'i), only a few are truly dangerous to humans. Spiders and insects have evolved together, and insects have largely been food for spiders. Despite the common perception, most spiders are shy and retiring and are not aggressive; bites to humans happen when the spiders are trying to protect themselves. In fact, many bites blamed on spiders are actually bites from insects or other arthropods—ticks, beetles, fleas, bed bugs, mites, and bees, for example.

Willis Gertsch in his informative book *American Spiders* wrote, "By far the majority of spiders are relatively helpless creatures, always willing to scurry out of the way, never attempting to bite without the greatest provocation. It must be remembered that a spider bite is always a pure accident."

The widow spiders are the only truly dangerous species in Hawai'i, but several other species are able to inflict painful bites and slow-to-heal wounds. Despite what you may have heard, Hawai'i does NOT have the notorious brown recluse spider. A relative called the brown violin spider does occur here, and though its bite can cause a nasty wound, it is not as bad as that of the brown recluse. All harmful spiders in Hawai'i were dealt with in *What Bit Me?* Only species normally encountered inside the house are mentioned here. For further information on harmful spiders, see *What Bit Me?*

Tarantulas (Fig. 89) do not occur in Hawai'i. The large brown, or cane, spider that comes into the house is often confused with a tarantula because of its size, but it is harmless and is a great cockroach hunter.

If you cannot stand spiders in the house under any circumstances, you can, of course, kill them. Or you can catch the spider (see below) and release it outside. As noted below under the widow spiders, however, we do

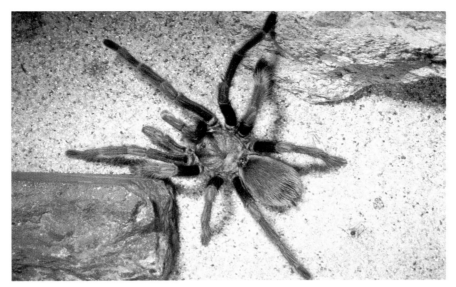

Figure 89. Tarantulas are heavy-bodied, thick-legged, and hairy. They DO NOT occur in Hawai'i. Many people mistake the cane spider for a tarantula. (Photo courtesy of Van Waters & Rogers Inc.)

NOT recommend putting up with these spiders because their bite can make you very sick.

Widow, Violin, and Cane Spiders

Southern Black Widow Spider
Latrodectus mactans (Fig. 90)
Brown Widow Spider
Latrodectus geometricus (Fig. 91)
Western Black Widow Spider
Latrodectus hesperus
Family Theridiidae

One species of widow spider is commonly encountered in Hawai'i and is found on all main islands: the brown widow, *Latrodectus geometricus*. The southern black widow, *Latrodectus mactans,* used to be quite common, but is rarely seen nowadays. The Western Black Widow, *Latrodectus hesperus*, is found on Moloka'i, O'ahu, Maui, and Midway. *Latrodectus hesperus* is similar in appearance and medical importance to the other widow spiders.

Widow spiders are found in the drier regions of all inhabited islands. They are secretive and make their webs in dark, out-of-the way places such

Figure 90. Widow spiders can be recognized by the red or orange hourglass on the underside of the abdomen. The southern black widow (here, with her smooth, white egg case) is not as common as it used to be in Hawai'i. (Photo courtesy of Van Waters & Rogers Inc.)

as basements, garages, storage sheds, crawl spaces, crevices, woodpiles, stacks of lumber or boxes, and hollow-tile blocks. Brown widows are more common in Hawai'i than black widows, but all three widows prefer dark, low-traffic locations and spin their webs near ground level.

Description: The southern black widow female is glossy black to dull brown, usually with a bright red, hourglass-shaped mark on her underside. The smaller male often has yellow markings at the sides of his abdomen. Females are 3/8" long, males 1/8". The brown widow female is usually mottled brown or gray with blackish, red, and yellow markings, occasionally entirely black, so that it may be confused with the black widow. The mark on the underside is dull orange or reddish and varies from an hourglass shape to two dots. The spider is similar in size to the black widow. Young spiderlings of widow spiders often have colorful patterns on the tops of their bodies.

The best way to tell the brown widow from the black widow is from the surface of the egg cases, which you will almost always see in webs with the spiders. A web may contain one to several egg sacs. In the southern black widow, the cream-colored egg sacs are rounded, smooth, and papery looking; in the brown widow, scattered tufts of silk give the brownish egg sacs a bumpy outline. The western black widow egg case is also smooth, but instead of being round like the southern black widow's, it is pear-shaped.

Evidence: The widow females spin a coarse, messy, tough web in a dark, sheltered place usually near ground level and feed on insects caught in the

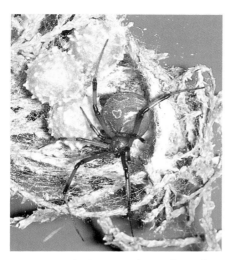

Figure 91. The brown widow is the widow spider most often seen and reported in Hawai'i. Note her brownish, bumpy egg case. (Photo by Larry Nakahara)

web. Widow spiders are active mostly at night. Females are shy and usually bite only in self-defense, usually when pressed on or crushed against something; males apparently do not bite or cannot penetrate the skin. The brown widow seems less likely to bite, or it injects less venom than the black widow.

Impact: The black widow is noted for its potent venom that is dangerous to humans. Two tiny red spots usually mark the entry of the widow's fangs. The bite usually feels like a pinprick. The venom is very strong, and the reaction to the bite depends on the amount injected and the susceptibility of the victim. Bites of black widows primarily affect the central nervous system (neurotoxin), but skin reactions can occur. An intense burning sensation and excruciating pain around the bite site may be felt, and heavy sweating or salivation and nausea and vomiting may follow. Severe cramps, rigid abdominal muscles, convulsions, and shock may also occur. Symptoms may appear immediately after the bite or up to several hours afterwards. In severe cases, the poison may also cause breathing problems, slurred speech, and paralysis. First aid other than antiseptic on the bite should not be given. If you are bitten, stay calm and seek medical treatment at once. A specific antivenin (Lyovac) is available and should be given as soon as possible.

Children and elderly persons with medical problems are at highest risk. Less than 1% of those bitten by black widows die, most of those from untreated bites.

Brown widow venom is potent, but only about one-fourth to one-tenth as strong as that of the southern black widow. However, if you are bitten, treat your condition as serious and get medical help quickly.

Prevention: If you have widow spiders in your area, teach your children how to recognize them and instruct them not to touch the webs and to let you know about them so that you can arrange for removal of spider and web. Remember that removing the spiders and webs may simply vacate the niche for another spider to move into. So after removing the spider, make changes to the area to make it less attractive to other spiders. Remove clutter, unstack lumber, increase the light, caulk crevices, remove piles of paper or boxes, and so forth. If you move things in potential widow areas, remember to use gloves and long sleeves to avoid being bitten.

Control: Though bites from widow spiders are rare and the spiders are excellent predators on other insects and small arthropods, because of their possible danger we think that they should not be allowed to remain around the house, especially if there are young children in the household. The soft, delicate spiders are easily killed by crushing with a broom or stick. Webs and eggs should be destroyed by removing and disposing of them or by vacuuming; the vacuum cleaner bag should be removed and put in a plastic bag for disposal.

Brown Violin Spider
Loxosceles rufescens
Family Loxoscelidae
(Fig. 92)

Figure 92. The faint outline of a violin on the carapace tells you that this is a brown violin spider. (Photo by Larry Nakahara)

This is a long-legged, yellowish spider whose bite may cause severe pain and a slow-to-heal gangrenous wound. The brown violin spider is a close relative of the infamous brown recluse *(Loxosceles reclusa)* (Fig. 93) of the southern and central United States, which does NOT occur in Hawai'i. Their habits are similar, although the venom of the species in Hawai'i seems to be less toxic. *Loxosceles rufescens* is widely distributed over the world. It is reported from O'ahu, Kaua'i, and Maui, has been found under old boards and loosened bark, and occasionally enters homes.

Description: The brown violin spider's carapace (top of the body behind the head area) is light yellowish brown tinged with reddish orange and carries a faint, violin-shaped mark (with the violin's neck pointing backward). Violin spiders have six eyes arranged in pairs in a semicircle surrounding the base of the violin; most spiders have eight eyes. The abdomen is pale brownish gray. The legs are fairly long. The females are about 5/16" long, the males about 1/4" long.

Figure 93. The brown recluse spider is NOT found in Hawai'i. The violin on the back is much sharper in outline than in Hawai'i'144s brown violin spider. (Photo courtesy of Van Waters & Rogers Inc.)

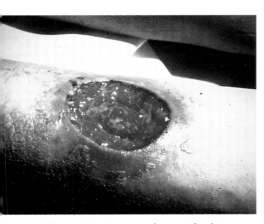

Figure 94. Brown recluse spider bites can cause surrounding tissue to die. Brown violin spider bites are not nearly as bad, although they may be painful and cause some tissue loss. (Photo courtesy of Van Waters & Rogers Inc.)

Evidence: Violin spiders live in dark places and spin large, irregular, sheetlike, bluish webs. Like the webs of widow spiders, the webs are messy and usually at ground level. These spiders are active at night and are not aggressive. Unlike the widow spiders, however, violin spiders sometimes hunt away from the web and may take temporary shelter in clothing, boxes, and similar places. Both males and females are venomous.

Impact: The venom of this spider contains a cell-destroying factor that may cause skin and surrounding tissue to die (Fig. 94). Bites are usually on arms and legs, and reactions may range from mild redness to serious tissue destruction. A slight stinging sensation is sometimes followed by intense pain. Often the victim is not aware of a bite until several hours later. Blistering, swelling, or reddening may occur around the bitten area. In a day or so, the skin may turn purple, followed in a week or so by blackening as the cells die. Tissue eventually sloughs away, sometimes leaving large pits in the skin. Healing may take several weeks. If not properly treated, gangrene may result. If bitten, clean the wound and apply antiseptic. Surface treatment of the bite may not stop additional damage to the skin, so it is important to see a doctor as soon as possible. Ice packs around the wound may keep the poison localized. No fatalities have been reported from the brown violin spider. No antivenin is available to counteract the venom.

Large Brown Spider
Heteropoda venatoria
Family Heteropodidae
Also known as Huntsman Spider, Banana Spider, Cane Spider
(Figs. 95, 96)

The large brown spider is widespread in the tropics, and in Hawai'i is found on all major islands. These spiders usually live outdoors under tree bark and refuse, but are frequently found in houses and other buildings hunting on walls, especially in wetter parts of the Islands. They are generally beneficial because they feed on many pest arthropods, including cockroaches in the house.

Description: This large, hairy-looking spider has a light brown to yellowish body often over an inch long in females (smaller in males) and long, bristly, outspread legs spanning 3" or more. Large brown spiders are often mistaken for tarantulas, which do not occur in Hawai'i. They do not spin webs. During egg-laying periods, a female may be seen carrying a large, white, pill-shaped egg sac in her mouthparts. She does not feed during this month-long period of protecting her eggs.

Figure 95. Have you ever had friends that others didn't like? Here's one, the large brown (or cane) spider. She's our good friend, capturing cockroaches and other bad insects, yet no one seems to like her. (Photo by Gordon Nishida)

Figure 96. Here, the cane spider is backed up into her defensive posture. She rarely bites unless forced to defend herself. (Photo by Gordon Nishida)

Evidence: Presence of a large, brown spider on walls.

Impact: Despite its ferocious appearance, this spider rarely bites unless teased to do so; if it bites, it may not inject poison. However, it has been reported as biting and affecting humans in other parts of the world. The bite may feel like a slight pinprick. Reactions to the bite are pain, followed by redness and swelling, and possibly a raised bump at the bite site. If you are bitten, take normal precautions and see a doctor promptly if problems develop.

Preferred Food: These spiders feed on insects and other arthropods, including cockroaches, flies, silverfish, and other household pests.

Preferred Locations: Large brown spiders hide in crevices during the day and come out at night, waiting on walls and ceilings and searching for food.

Prevention and Control: Not necessary.

Other Spiders

A few other spiders found around the house that may bite were covered in *What Bit Me?* These include the pale leaf spider *(Cheiracanthium diversum)* (Fig. 97) and the Asian spinybacked spider *(Gasteracantha mammosa)* (Fig. 98). Other spiders frequently found in the house are relatively harmless. We suggest tolerating them—even welcoming them. Following are examples of some common household spiders in Hawai'i.

Family Pholcidae. Some of the most conspicuous of the small web spinners are the pholcid spiders. These spin loose, irregular webs in dark places. *Pholcus phalangioides* (Fig. 99) is found throughout the world and is common in homes in Hawai'i. Its elongate body (about ¼") and exceedingly long legs (about 2" long) cause people to mistake it for a daddy longlegs (harvestman), which is not a spider at all, but a spider relative. Webs can be

Figure 97. The pale leaf spider has very large fangs and can give a painful bite. (Photo by Ron Heu)

found almost anywhere in the house. We have even seen them in shower stalls that have not been used for a few weeks. After a while, their messy webs become littered with bodies of prey; if this bothers you, simply destroy the web without killing the spider and she will spin a fresh web. If disturbed, this spider often starts bouncing back and forth, shaking the web, probably to discourage predators.

Figure 98. The Asian spinybacked spider makes large webs, usually outside on plants. It can bite. (Photo courtesy of Van Waters & Rogers Inc.)

Family Theridiidae. *Theridion rufipes* (Fig. 100) is a bulbous, reddish brown spider in the same family as the widow spiders, but is much smaller in size. There are unconfirmed reports that this spider bites, but we have no firm evidence at this time. These spiders hang upside down in their webs usually spun at the base of walls in corners and angles. They feed on insects and other arthropods that become trapped in

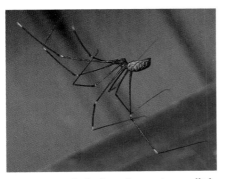
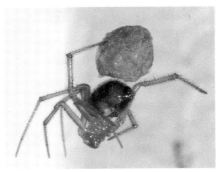

Figure 99. Pholcid spiders spin smallish webs, usually close to ground level. This spider has distinctive white bands at the ends of the leg segments. Pholcids are excellent predators of insects. (Photo by Gordon Nishida)

Figure 100. *Theridion rufipes* is often found in homes, spinning its webs in corners and angles of walls. Usually attached to the webs are reddish brown egg cases slightly larger than the spider itself. (Photo by Gordon Nishida)

their webs. If you can put up with their messy webs, they can be very useful in eliminating pests in your home.

Family Salticidae. Salticids are tiny to small, brownish to blackish, large-eyed jumping spiders that hunt freely around the house during the daytime. These friendly little creatures will often perch on a hand or finger. They have good eyesight and are able to spot prey at a distance, creep up on them, and pounce like a cat. They have a short, stout, furry body with shortish legs. They run, jump, hop, and leap from place to place and are fun to watch. They are also helpful in keeping down insect populations. Don't kill these cute little guys and gals!

Figure 101. The daring jumping spider is the largest jumping spider known to be established in Hawai'i. Other jumping spiders are smaller and usually brown in background color. (Photo courtesy of Van Waters & Rogers Inc.)

Hawai'i has one large jumping spider that is known to bite humans. The daring jumping spider, *Phidippus audax* (Fig. 101) is relatively large (males $1/4-1/2$" long and females $5/16-5/8$" long), short-legged, and furry. The abdomen is black with a white crossband and a few spots. The fangs are greenish. The spider is known from O'ahu and Maui and occasionally enters homes. *Phidippus* bites are sharp and painful and produce pale, raised bumps surrounded by redness,

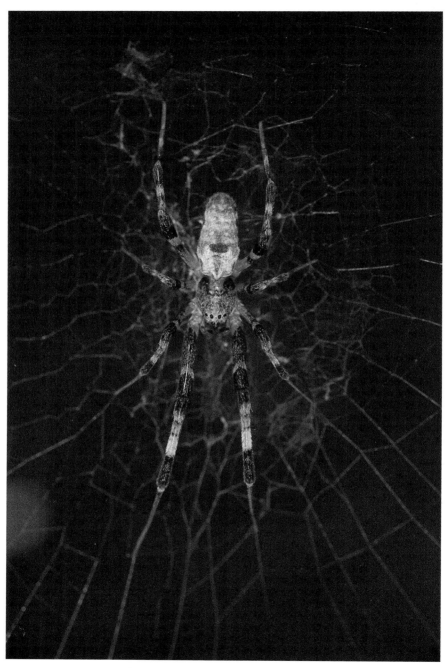

Figure 102. *Zosis geniculatus* prefers making its large webs in darker places, under the house, in corners of lānai, in garages, and the like. (Photo by Gordon Nishida)

accompanied by blistering and swelling. The swelling can be severe, extending beyond the immediate bitten area, but usually subsides within 48 hours. A dull throbbing pain and itchiness may last several days. Treatment is usually not necessary other than with an over-the-counter pain remedy. If symptoms are severe or if complications develop, see a doctor.

Family Uloboridae. The uloborid spider *Zosis geniculatus* (Fig. 102) likes darker places, especially under the house. It is very slender, grayish white, and spins large webs in which it hangs with the legs extended forward and backward. This behavior makes the spider look like merely a thickening along one of the webs. The webs are not as sticky as some, but very tough and stringy.

You might also find in your house the house spider *(Tegeneria domestica)* and spitting spiders of the genus *Scytodes*.

Capturing/Identifying Spiders

If you think a spider has bitten you, take it in for identification. To capture a spider alive for identification, find a jar large enough to hold the spider. Place the jar over the spider. Slide an index card or cardboard beneath the mouth, capturing the spider inside; turn the jar over and cover, making some holes in the cover so the spider can breathe. If you don't want to keep the spider alive, put the jar in the freezer for a few days. Or you can put rubbing alcohol in the jar (don't make holes in the lid in this case) to cover the spider and preserve it until you can take it for identification.

Control of Spiders

Web-building species build their webs where insects will be most likely to blunder into them. Spiders are the original pest-control specialists! If webs are dusty and dirty (cobwebs), they are often no longer in use. Vacuum them up. If the spider is still active in the web but the web is unsightly from the corpses of tiny victims, you can clean up the web without hurting the spider—leaving the spider to spin a new web and live on to do more pest control.

The following measures will be helpful in keeping down populations of harmful spiders and help you to avoid bites:

✔ Do not allow refuse, which supplies homes for widows or other harmful spiders, to build up in the yard.

✔ Use gloves and long-sleeved garments when cleaning out items in basements, storage closets, and wood and lumber piles.

✔ Shake out items that have not been used for a while before wearing or using them; items like clothes, shoes, and blankets kept in long storage are good hiding places for secretive spiders.

✔ Seal containers like boxes and other storage containers so that spiders and insects can't get inside.

✔ Thoroughly vacuum corners, baseboard areas, behind furniture, and similar places.

✔ After removing widow spiders, change the habitat so that another spider does not take its place.

SPRINGTAILS
Order Collembola

170 species in Hawai'i, 95 native

Springtails (Collembola) are tiny little animals, less than 1/5" long, usually light colored, often white. They are called "springtails" for the forked muscular appendage at the end of the abdomen, which is used for springing into the air. If you disturb this animal, it will usually jump away. Springtails occasionally invade homes, especially in areas of high humidity such as basements, bathrooms, and kitchens. They prefer darker areas and avoid light.

Springtails feed on algae, fungi, fungus spores, pollen, and decaying vegetable matter; they do little or no damage in the home.

If you have problems with springtails, remove sources of high humidity, such as moist leaves and other damp and moldy materials in rooms or near the foundation of the house. In very moist locations, a dehumidifier may help. These insects are not as much of a problem here as in mainland areas.

TERMITES
Order Isoptera

5 species in Hawai'i, 0 native

Termites are the most destructive and economically important of all pest insects. Damage to homes in the United States is in the billions of dollars each year. Ironically, termites are only doing what they've always done—eat wood—and our housing design practices have provided them with plenty of opportunities and lots of food sources. They are probably the only insects that can digest the cellulose in wood, thanks to the friendly protozoans that live in their intestines.

In forests, termites are useful in eliminating dead trees and debris and in recycling nutrients. Unfortunately, in replacing forests and trees with our own wooden shelters, we have offered up our homes to replace the termite's natural foods. We've created high-density "forests" of dead trees—our houses—for them to eat.

Termites are widely spread across almost the entire United States. Two types of termites are important in damaging wooden structures: drywood termites and subterranean termites. Subterranean termites are by far the most damaging, and in Hawai'i, the Formosan subterranean termite has been the major insect pest for decades, causing staggering amounts of damage—about $100 million annually.

In contrast to the detection and control of other pests discussed in this book, successful termite management usually requires some professional help. However, the more that we as homeowners know about termites, their biology, what conditions are favorable to them, how to detect their presence, and what to do when we find them in our homes, the better we can fight this destructive pest. For subterranean termites, we must act quickly, but we still have time to make informed decisions; don't be so hasty to rid your home of termites that you make ill-advised decisions that may be more costly in the long run.

Much information in this chapter is summarized from publications and teaching materials of the College of Tropical Agriculture and Human Resources, University of Hawai'i, and from the research of Dr. Julian Yates III, Dr. Minoru Tamashiro, and their colleagues. This group has worked for many years on termites, conducting field tests for useful control methods under Hawai'i conditions. This is important because what works for termites in the mainland United States and other parts of the world may not work well in Hawai'i because of differences in temperature, humidity, weather conditions, and soil types.

Termites versus Ants

It is important that you distinguish between ants and termites because the measures for their control are different. Termites are often called "white ants," but they are not related to ants. Because the winged adult termite is the only stage that leaves its nest, you will most likely see the winged stage, which does look somewhat like some ants, especially winged ants. Here are the major differences that you can see with the naked eye: ants have a narrow "waist," termites are thick in the waist; ant antennae are bent or "elbowed," whereas termite antennae are straight; termite wings are longer, about twice as long as the body, and the two pairs of wings are about equal in length; ant wings are shorter, with the hind pair much shorter than the front pair. If the wings are broken off, termite wings leave little stubs on their backs; ant wings break off cleanly.

Compare the insects shown in Fig. 103 and the wings shown in Diagram 2 on page 41.

Figure 103. A winged termite *(left)* and a winged ant *(right)* are shown to help you tell them apart. When the wings are in place, it's difficult to tell if the waist is thick or narrow, but you can see the wing venation differences (see page 41) (Photos by Gordon Nishida and Van Waters & Rogers Inc.).

A Mutually Beneficial Relationship: Symbiosis

Symbiosis describes an association in which two organisms live together intimately and both benefit from the relationship. And so it is with termites. Termites cannot by themselves digest wood. Luckily for them (and unluckily for us), they harbor in their gut one-celled protozoans that break down cellulose and make it digestible by the termite. Termites are born without the protozoans. Until their third stage of development, they must be fed predigested food by adults and older nymphs. By the third stage, the protozoans have been transferred to the young termites in their food and by grooming, and the young are finally able to digest the cellulose in wood.

Swarms

Major swarms of termites occur from May to July, with smaller swarms any time of the year. This is the way that termites spread to a new area. Termites are poor flyers and usually stay within a quarter mile radius of the emergence site. Swarms begin after sundown on warm, humid, still evenings. If the wind speed is more than 2 miles per hour, swarms do not start or they stop.

Adults emerge from their nests and fly for 10–30 minutes before falling to the ground. They drop their wings, and males and females pair off. Pairs can be seen running around in tandem, the male behind the female. Most termites are most vulnerable in this swarming stage, and most are eaten by predators or die from exposure or failure to find food or a nest site. See sections below on drywood and subterranean termites for additional information.

What Termites Do I Have?

It is important to know which termite is in your home because treatment for drywood termites is different than that for subterranean. Moreover, subterranean infestations are much larger, potentially much more expensive, and may affect the very structure of your home. You will need to treat quickly if you have subterraneans.

Drywood termites live in much smaller colonies and build nests inside wood, away from the soil. Because they don't have high requirements for moisture, they can be transported easily in lumber, furniture, cardboard, or wooden objects. They eject fecal pellets from wood (Fig. 104) they are nesting in through small, round holes, which they then plug up. These are called "kickout" holes (Fig. 105), and the pellets and holes are signs of drywood termites. The pellets range in color from light brown to black depending on the type of wood eaten. Piles of these pellets suggest that a termite infestation is probably active in the wood directly above or beside the holes.

Figure 104. Drywood termite frass (droppings) can be light to dark-colored, often depending on what wood the termites are feeding on. (Photo by Gordon Nishida)

Figure 105. Piles of drywood termite frass are found beneath kickout holes. These "termite toilet" openings can be seen to the left of the dime. (Photo by Julian Yates)

Subterranean termites in Hawai'i are among the most aggressive of all termites. This termite also eats a wide variety of woods and cellulose materials. It also can penetrate nonorganic materials such as soft metals and cracked concrete to reach food or water. It does lots of damage, quickly, and can cover lots of land with its colonies—as much as an acre. It can do extensive damage if undetected. A home of unprotected lumber and with no soil treatment (chemical or physical) built over an existing flourishing colony of termites can be completely destroyed in a few years in Hawai'i. Conservative estimates of costs to prevent, control, and repair damage by this pest are around $100 million a year in Hawai'i and these costs will continue to increase.

The great danger of the subterranean termite is that it does its damage while unobserved. There are often no visible signs of its presence until a door sags, a roof leaks, paint blisters, wood floors become soft or springy, or electrical problems develop. Preventive measures, early detection, and quick treatment are critical for dealing with this pest.

Drywood Termites
West Indian Termite
Cryptotermes brevis (Fig. 106)
Lowland Tree Termite
Incisitermes immigrans
Forest Tree Termite
Neotermes connexus

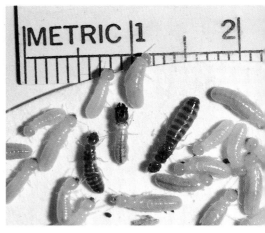

Three drywood species occur in Hawai'i. Only the West Indian drywood termite is of economic importance. This is the species you may see in your house, so we will limit our discussion to this species.

Figure 106. The king and queen of the West Indian drywood termite are darker in color (in center, separated by a soldier and surrounded by workers). The queen is the larger one on the right. (Photo by Julian Yates)

Cryptotermes brevis, probably introduced from tropical America before 1869, is now found on all the major islands. It prefers hardwoods, so it is often found in expensive furniture. It also likes books and other materials containing cellulose.

Swarms begin as winged adults that emerge from kickout holes. Adults pair off and look for a site to start a colony. Any hole, crack, or open joint may provide access to a suitable site containing cellulose. Drywood termites obtain water from normal humidity and the food they eat, so they do not need high-moisture conditions.

After a site is found, the adults seal themselves in and mate. The colony is very slow-growing. Small numbers of pink, jellybean–like eggs are laid shortly after mating and take perhaps 2 months to hatch under summer temperatures. A typical year-old colony is small, and mature colonies rarely exceed 300 individuals. Larger numbers may be found in heavily infested wood where colonies have merged. Because the colonies are usually relatively small compared with the thousands and even millions of individuals in a subterranean colony, it takes a long time to reach a size that can do significant damage. Thus, there is a good chance that a homeowner can eliminate drywood termite colonies before they become very large.

Figure 107. The single greatest threat to wooden structures in Hawai'i is the Formosan subterranean termite. A dark-headed soldier and some workers are shown here. (Photo courtesy of Van Waters & Rogers Inc.)

Subterranean Termites
Formosan Subterranean Termite
Coptotermes formosanus
(Fig. 107)

This termite was probably already present in Hawai'i before 1869, apparently introduced from Formosa or South China during the period of sandalwood trading between Hawai'i and China. As of 1995, it is on all the major islands, but is most widespread on O'ahu and Kaua'i, with spotty distribution on the other main islands.

After pairing off and sealing themselves in the mating chamber, male and females mate and lay a batch of 15–25 eggs within 5 days. Eggs hatch in less than a month. Like drywood termites, newborns must be fed by other termites until they have acquired the protozoans after two molts and can be functional workers in the colony. The queen lays another batch of eggs and the process continues. A 7-year-old colony may total 2 million termites, and a large colony may contain 10 million individuals (Fig. 108).

The mature queen, no more than a massive egg-laying machine (like the giant alien mother in the movie *Aliens*), can no longer walk and must be fed and moved by workers. At this gross stage she can lay about 2,000 eggs each day. She and her king, who produces the sperm to fertilize her eggs, may live 20 years or more. You can see how the large number of eggs produced quickly in a long-lived queen can make this termite species so dangerous.

Soldiers have large, brown heads, with pincerlike jaws used for fighting; they cannot feed themselves and must be fed by the workers.

Supplementary queens and kings exist in the colony as well. They are wingless, blind, and not as strongly colored as the primary reproductives that swarm and form colonies. The supplementaries never leave the colony. Their function is to take over reproduction if the primary king or queen dies or becomes separated from the main colony. The presence of these supplementaries makes controlling subterranean termites much more difficult. If the soil colonies are cut off from the part of the colony already in the house through insecticidal treatments, supplementaries can take over reproduction and allow the infestation to continue as an aerial colony. Even without supplementaries in the house, workers can live 4 or more years and can by themselves do considerable damage before they die out. Thus, the infestation in the house must be treated as well.

Figure 108. Formosan subterranean termites are dangerous because given adequate food and water their populations can grow to massive size in a short period of time. (Photo by Julian Yates)

Tunnels. Workers are the workhorses of the colony. They are white, blind, and forage for food, feed the king and queen, take care of the eggs and young, build tunnels, and generally maintain the colony. They live for 4–5 years. Because they dry out easily, they work within earthen tunnels and galleries. The worker stage is responsible for all the damage caused by termites.

Workers can build their earthen tunnels over concrete foundation walls up to wood, which they then enter through cracks. They also build tunnels over wood to bypass treated wood. Be alert for these tunnels around the outside of your buildings. They are one of the few visible signs you will have of a termite infestation. If you see tubes, knock them apart. If the tubes are rebuilt within a few days, you know that there is termite activity there.

Inviting Conditions for Termites

Many features of buildings and landscaping provide favorable conditions for termite entry and development. The following are a few of the architectural trouble spots:

✔ allowing wood to come in contact with the soil

✔ improper installation of termite shields

✔ sprinklers installed close to walls

✔ gutters, drains, and downspouts near walls; water must be carried away from the house

✔ improper soil grade; water should properly drain away from the house

✔ wooden fences in contact with structure

✔ use of untreated wood

✔ soil backfill against retaining walls

✔ backfill with wood debris

✔ shake roofs

Be sure you work with a contractor who is aware of termite prevention and control.

Detection of Termites

Figure 109. Damage (on left) by Formosan subterranean termites to lead-shielded cable. (Photo by Julian Yates)

Be vigilant around your home. Do one thorough inspection of the house at least twice each year, with professional help if necessary. Check for tunnels around the outside of the house.

Inside the house, check wood flooring, moldings around windows and doors, window sills, moldings along floors with wall-to-wall carpeting for any evidence of rippling or darkening. Tap the wood with an ice pick or screwdriver and listen for a hollow sound; if termites have

Figure 110. Damage by Formosan subterranean termites to a book at an O'ahu elementary school. (Photo by Julian Yates)

been feeding within, an ice pick will go easily through the skin of the wood, exposing the galleries within the wood. The wood strips under the edges of wall-to-wall carpeting that hold the carpet in place are a great delicacy for termites; these strips can be spot checked in suspicious areas. Be aware that materials other than wood can be damaged by termites (Figs. 109–111).

Fiberoptic scopes, moisture meters, and listening devices are sometimes used by professionals looking for termite infestations.

Figure 111. Formosan subterranean termites will feed on any paper products they can find. (Photo by Julian Yates)

Prevention of Termite Invasions

There's not a lot we can do to prevent an aggressive attack by subterranean termites. The first line of defense is to build structures protected by a termite barrier beneath the foundation; this provides permanent protection for your home (see Basaltic Termite Barrier and Termi-Mesh below). The second line of defense is to use only wood treated with one of several chemicals (see Wood Treatment below). Even if you live in a house that was not built with treated wood, at least you should use treated wood to build additions and structural components or to make repairs.

Homeowner's Termite Prevention Checklist

Homeowners must be ever alert against conditions that may encourage termite attraction and infestation. Some simple preventive measures can help protect your house.

✔ If you see termites beginning to swarm, turn off inside and outside lights to avoid attracting them to your area.

✔ If you see swarming termites inside your house, find how they are getting in and make repairs or adjustments.

✔ Kill any pairs you see running around together.

✔ Do not plant vegetation so close to the base of the house that the slab is screened from view. Cut down such existing plants so that you can inspect the base of the slab periodically for tunnels. Plants near the house also provide moisture and food for termites.

✔ Avoid having any wooden parts of the house come directly in contact with the ground. This provides a direct access for termites into your structure.

✔ Be sure that any wood or cellulose materials you bring into your house for construction or storage purposes are not infested with termites or have been treated.

✔ Keep the exterior paint on wooden structures in good condition. Deteriorated paint allows easy entry for termites.

✔ Wooden planters next to the house should not contact the soil and should not be in contact with the house structure.

✔ Keep wood, building materials, firewood, stumps, plant materials, and other such materials away from the house. Avoid storing these for long periods or store them on a concrete slab.

✔ Patch or repair cracks in concrete. These cracks give termites access to your house.

✔ Repair leaking pipes and faucets. These can keep soil under your house continually moist and invite termites.

✔ Inspect wooden fences periodically, especially the posts or parts that touch the soil, to make sure that they are not infested.

✔ Provide inspection access at plumbing intrusions (for example, cold/hot water pipes, bath traps, and such).

Wood Treatment

Two different companies offer Osmose and Wolman pressure-treated wood. The active ingredient is chromated copper arsenate (CCA), which turns treated wood a greenish color. Douglas fir wood treated in Hawai'i with CCA cannot withstand a sustained attack by an established Formosan subterranean termite colony because the chemical does not penetrate well into the wood and after the termites have eaten beneath the surface, there is little to turn them away. However, treatment with CCA is probably a deterrent to West Indian drywood termites and Formosan adults looking for sites to establish colonies.

Another substance called ACZA (ammoniated copper zinc arsenate) is extremely effective even against the Formosan termite. It has good penetration into the wood. Not only does it turn away Formosan termites under Hawaiian conditions, but it kills them as well. This chemical turns wood blackish, and wood treated in this way is more expensive than Osmose and Wolman.

Products containing sodium borate are also available in Hawai'i to dip, spray, or treat wood. They do not initially penetrate the wood thoroughly, but enough of the surface chemical is eaten by the termites to kill them rapidly. Borates continue to penetrate the wood after treatment and are said to penetrate 100% over time. The advantage of borates is that they do not discolor wood. However, these chemicals are water soluble and should not be used where exposed to rain or moisture.

A homeowner should not look only at immediate costs, but should also carefully consider effectiveness of treatments against subterranean termites. The most important factor is effectiveness. If you prorate the cost of initial preventive measures over the cost of the home, you will find that the additional expense for more effective treatments costs only pennies more viewed over the life of your home, especially when you look at possible destruction of the structure of your home and the cost of repair or replacement.

Shake Roofs. These are very popular in Hawai'i because they are attractive. However, they require high maintenance to keep them in good condition. They are frequently infested by drywood termites. If shakes are not properly maintained and they absorb moisture over time, subterranean termites may become a problem as well. If you already have a shake roof, it is wise to inspect shakes frequently and perform the necessary maintenance. If you are presently considering what roofing materials to use for your home, something other than shakes might be a better choice in Hawai'i.

Basaltic Termite Barrier (BTB) and Termi-Mesh

In the late 1980s, Walter Ebeling of UCLA showed that sand barriers of a particular particle size placed around the inside, outside, or on both sides of existing foundations protected buildings from termite attack. A layer of sand in the critical size range placed on the soil before laying of concrete foundations also gave protection. After tests conducted by Minoru Tamashiro and his group at the University of Hawai'i confirmed Ebeling's findings, the technique was written into the Honolulu building code in 1989. Commercial applications of BTB are now available. The idea is simple: underlay the foundation of a new structure with a barrier made of particles small enough that the termites cannot penetrate. The materials are inert and do not break down with time. What you have, then, is a safe, non-chemical preventive technique that stops termites but does not kill them, a preventive that lasts over the life of your home. BTB has proven very effective; unfortunately, it is not practical to install around older homes and is used mostly for new construction.

Termi-Mesh is also a physical barrier but is made of a stainless-steel screen material. It is a new product from Australia that has proven effective there. Hawai'i is the first U.S. franchise for the product.

Chemical Control of Termites

Drywood Termites

Step one is to determine how serious the problem is. You basically have three options: spot treat; fumigate only infested objects, such as furniture; or fumigate the entire house.

If infestation is restricted to a few areas of the house or is in the furniture, the areas can be spot treated. If the infestation is more extensive or spread throughout the structure, fumigation may be necessary.

Spot Treatment. After galleries have been located, an insecticide can be injected. Term-out is an effective aerosol formulation readily available in Hawai'i. It contains Resmethrin, which is odorless and relatively nontoxic to humans and pets.

Object Fumigation. If only furniture is infested, you may be able to fumigate only the furniture; several companies are equipped to do this kind of vault fumigation.

House Fumigation. If a professional has confirmed that your infestation is extensive, fumigation of the entire structure may be necessary. This is expensive and should be used only after you have had a professional inspection to confirm that it is needed. Fumigation is not done to keep termites out; the chemical has no lasting effect. It is only used when there is an extensive active infestation of drywood termites. Fumigation is not effective for subterranean termites in the soil, because the colonies cannot be reached by the fumigation gas. However, if subterranean termites have invaded the house extensively, soil treatment and fumigation may both be necessary.

Fumigation means covering the entire structure within plastic-coated nylon tenting materials and sealing the tent against escape of gas. Care must be taken to be sure that no living things are left in the structure and that food products are properly sealed. Vikane (sulfuryl fluoride) used in fumigation is effective against several household pests besides termites and will kill cockroaches, ants, and pantry pests too. Before Vikane is applied, chloropicrin (tear gas) is pumped into the tent to alert any person that might have been left behind accidentally. Vikane is then pumped in and left in at least overnight.

Fumigation kills only insects in the house at the time of treatment. It has no residual effect and does not prevent reinfestation.

Subterranean Termites: Chemical Treatment of the Soil

Before the 1980s, chlordane and other chlorinated hydrocarbon insecticides were used for soil treatments in Hawai'i against the Formosan subterranean termite. Chlordane is a high vapor-pressure termiticide, which gives good penetration in soil and is highly toxic to termites. It can be applied fairly haphazardly without regard to soil type or moisture and still kill; its period of effectiveness is 20–30 years. Chlordane was banned in the United States in April 1988. The loss of chlordane has meant that we carefully consider many factors before deciding on a chemical for treatment. We no longer have the luxury to do sloppy or ineffective applications.

Several chemicals are now in use in Hawai'i, most of them repellents. They are all also poisonous, so they combine both properties of repelling and killing. However, they do not have the staying power of chlordane; they last a relatively short time.

The University of Hawai'i group at the College of Tropical Agriculture and Human Resources continues to conduct tests on the longevity and effectiveness of termiticides in Hawai'i. Results reported by the USDA Forest Service from tests conducted on the U.S. mainland are often not meaningful for Hawai'i conditions. Our soil and climatic conditions are very different from those of the areas where the standard USDA and EPA-sanctioned data are gathered. Also, Hawai'i tests are conducted using the Formosan termite and take into consideration its differing foraging habits. Criteria measured in the Hawai'i tests include soil type, rainfall, depth of penetration of the chemical, ability of the chemical to stop (repel) termites, and others.

In Hawai'i's tropical environment, soil insecticides do not remain effective for many years. New insecticides become available every few years and older ones decline in favor. You need to be current on the best and most effective chemical for your area. If you must have soil treated for subterranean termites, your pest control operator has access to the latest reports of termiticides tested by the University of Hawai'i group. There are also several guidelines for treatment of new building sites to avoid future problems with termites. Make sure your contractor seems knowledgeable in these areas. For example, clay soil, rainy areas, and wet soils present particular problems in Hawai'i, and termiticides used under these conditions must be chosen carefully .

Other Methods of Detection and Control of Termites

Tadd Dogs

Tadd Dogs (Fig. 112) puts pest inspection in the paws (and noses and ears) of canines. Tadd Dogs are trained to detect wood-damaging insects, particularly termites. These dogs, usually beagles, can squeeze into places humans can't reach. They are trained to smell termites and hear termite activity. The advantage of these dogs is that they find only living insects, not dead ones that may have been destroyed in a previous control campaign. They also key on the area infested, so that treatment can be localized. Dog-assisted inspections are admissible in court when claims are made against guarantees issued by pest-control companies. Tadd Dogs operates in Hawai'i under the license of pest-control companies licensed to do business in Hawai'i.

Figure 112. Trained dogs are now being used to find active termite infestations. (Photo by Julian Yates)

It is said that Tadd Dogs are very good compared with other methods of detection. If they key on a potential infestation, small holes can be drilled in a wall for inspection by a fiber-optic scope, which allows confirmation of an infestation behind drywall and paneling before further action is undertaken. Tadd Dogs do occasionally make mistakes; sometimes they are misled by another insect, such as a beetle inside a wall, and we have been told that once in a while they are dead wrong. Tadd Dogs carries Errors and Omissions Insurance in case the dogs are wrong and damage results from the erroneous detection. The service is expensive.

Electrogun

This device has been around for almost a decade and has been used successfully in parts of the U.S. mainland for treatment of drywood termites and wood-boring beetles. The operator sends a high-frequency, low-voltage current into the termite galleries. Because the galleries and termites are slightly moist, the electricity travels along the galleries, killing the termites without harming the wood. It has been shown to be safe and effective for localized treatment. However, in 1995 it had not yet been approved for commercial use in Hawai'i. The Electrogun apparently requires operators trained in the method. Some say that the potential is limited only by expertise of the operator; others believe the treatment is only appropriate part of the time. The gun will not work if the infested area is not accessible, nor if there is a large amount of metal in the structure. It is also not as effective as other methods if the infestation is widespread.

Heat

This is a promising new method for treating insect infestations, including termites. The process uses propane heaters and fans that circulate 120–150°F heat throughout a tented structure. Insects cannot live more than 20 minutes in 120°F heat, but higher heat is necessary to reach insects deep inside wood. Obviously many sensitive appliances, equipment, and foodstuffs in the house that cannot withstand high heat must be removed before treatment.

This is a friendly alternative to pumping toxic chemicals into a house, but the process is labor-intensive and costs more than chemical fumigation. Thermal eradication has just recently been introduced into Hawai'i, and we will wait to see how effective the method is here.

Bait System

A new approach to killing termites is undergoing field tests in Hawai'i. DowElanco's Sentricon system involves a wood product as bait, combined with an insect growth regulator as poison. The bait stations (wooden stakes), consisting of an openwork frame trap, are pounded into the ground in areas around a structure. Poison is then added; bait and poison are easily removed and replaced through a cap at ground level. Termites attracted to the wood pick up the poison and carry it back to the colony. As the slow-acting poison is consumed by colony workers, the development of the colony is disrupted.

The advantage of the Sentricon system is that it works for controlling termites in both new construction and existing sites.

Biological Control of Termites

Ants. Ants are among the most important termite enemies. They can attack exposed termites or swarming termites and kill them in great numbers. Breaking open earthen tunnels will help ants access termite colonies.

Diseases. Long-term tests continue on insect diseases that can be introduced into termite colonies by foraging workers, who might pick them up at bait stations. Termites live in a confined, humid environment that is favorable for fungus growth. Two promising fungi are presently under study by researchers in Hawai'i.

TICKS
Class Arachnida: Subclass Acari

9 species in Hawai'i, 1 native

Ticks are generally parasites of wild animals but also infest domestic mammals and birds. They harm humans and animals by feeding on their blood, injecting toxins, and transmitting organisms causing diseases. Of the nine tick species in Hawai'i, only the brown dog tick is a pest in the house and yard. It bites humans infrequently and is not known to transmit disease microorganisms in Hawai'i. More about the medical aspects of ticks in Hawai'i can be found in *What Bit Me?*

Brown Dog Tick
Rhipicephalus sanguineus
(Fig. 113)

The brown dog tick is found throughout warmer parts of the world and is probably the most common tick in the world. In Hawai'i, it has been reported from Kaua'i, O'ahu, Maui, and Hawai'i, but it is likely to be on any island that has a dog population. This common parasite of dogs can become a serious pest in the house in Hawai'i. It can build up large populations in a short time when pet dogs are present in the house.

Description: Before feeding, males and females are about 1/8" long, flattened, and reddish brown. After a blood meal, the female may enlarge up to 7/16 – 1/2" in length and 1/8" in width, and change in color to a grayish blue or olive green. Males do not expand as much after feeding.

Evidence: On a dog, relatively small ticks can often be seen clustered around a huge, engorged female; these are not "babies" but rather males. Larvae, or seed ticks, are six-legged and tiny before feeding; they are easy to miss when searching a dog for ticks. This tick likes to crawl upward and if the infestation is heavy, immature stages and adults may be seen on walls or ceilings, looking for places to molt or lay eggs. Larvae and nymphs can survive for many months without food or water, and adults can survive for almost 2 years without feeding.

Figure 113. The brown dog tick may be found in the house, especially if dogs live in the house. Shown are male *(left)*, female *(center)*, and female after feeding *(right)*. (Photo courtesy of Van Waters & Rogers Inc.)

Impact: Ticks feed by inserting their mouthparts (proboscis) into the skin of their host and sucking blood. The proboscis is armed with rows of backward-pointing barbs to help the tick stay in place during feeding, and this is why ticks are so difficult to remove after they have started feeding. After feeding and mating on the dog, adult females lay their eggs in cracks and crevices of the kennel or parts of the house frequented by the dog; as many as 5,000 eggs may be in an egg mass. The eggs hatch 3–12 weeks later into six-legged, light brown larvae. These seek a dog to feed on and spend 3–6 days sucking blood; then they drop off the dog and, in 6–23 days, molt into eight-legged nymphs. These then attach to a dog, feed for about a week, drop to the ground, seek a sheltered area, and in 2–4 weeks molt into an adult. Adults find a dog and feed for 1–8 weeks. You can see that if you have a few dogs in the house, they may be constantly attacked and fed on by ticks. Heavily infested dogs will become unhealthy and develop anemia. Dogs do not "catch" ticks by direct contact with infested dogs; ticks must first drop to the ground, molt, and seek another (or the same) dog for feeding.

Preferred Food: Blood. In urban areas this tick is usually associated with dogs, but where dogs are scarce, other animals such as rats or humans may be bitten.

Preferred Locations: Ticks attach around the head and neck and between the toes, areas where the tick is least disturbed by the scratching dog.

Detection and Monitoring of Ticks

Many of the same techniques used for fleas can be used for detecting and monitoring ticks. Review the section on fleas if you are having problems with ticks.

Using the same routine of combing your dog regularly for fleas will uncover ticks. Look especially behind the ears, on the neck, and between the toes. Do not try to remove ticks with a comb or brush, because the mouthparts may break off in your pet's skin. To remove a tick, apply tweezers to the tick as close as possible to the skin and pull gently but firmly, applying increasing pressure. Do not twist; pull straight out. Drown the tick in soapy water or flush it down the toilet. You may want to apply alcohol, peroxide, or an antiseptic to the bite wound to help prevent infection.

Control of Ticks

Controlling the brown dog tick in Hawai'i requires constant vigilance. The warm, humid weather year-round speeds up the life cycle and provides ideal opportunities for population explosions. A tick invasion is even more frustrating for the homeowner than fleas because of the often overlapping life cycles and the need for extreme persistence in applying control measures.

Cracks and crevices in walls, ceilings, floors, furniture, and such places in the dog's environment harbor egg-laying and resting ticks. These should be eliminated whenever possible in doghouses and kennels by sealing and caulking and/or treating with sorbtive dusts (see flea section). Vacuuming frequently in the dog's resting and sleeping areas will help reduce tick populations.

Stronger, more residual insecticides must be used to control ticks. Pyrethrum and pyrethrins do kill ticks and may be used in shampoos for dogs, but these insecticides do not last long enough to provide more than a few days' protection for the pet. Moreover, if you have a tick problem, you must treat your pet, the inside of the home if the pet comes indoors (including carpets and upholstered furniture), kennels, doghouses, and the yard. Unless these are all treated at the same time, your dog will be reinfested. Even with a complete treatment of all pets and all areas, some ticks are likely to survive and reinfest your pet. You may need to repeat the entire treatment in this case.

Organophosphate and carbamate insecticides, as well as insect growth regulators, will work on ticks. Consult your veterinarian for recommended dips for your pet and read insecticidal labels for those formulated for use inside the house and in the yard or kennel. In case of heavy infestations, we recommend that you consult pest-control professionals, who are experienced with stronger chemicals and have knowledge of when and where they are appropriate and safe to use.

WASPS AND HONEY BEES
Order Hymenoptera

Of all the plagues that heaven has sent,
A Wasp is most impertinent
— JOHN GAY

1,067 species in Hawai'i, 592 native

Most wasps are usually found outside, but a few occasionally stray inside. Some come inside to hunt for their prey. Wasps are very efficient predators and parasites of other insects and are mostly beneficial. Some wasps can sting, and care should be taken around them.

Wasps

Emerald Cockroach Wasp
Ampulex compressa
Family Sphecidae
(Fig. 114)

This beautiful wasp is also known as the jewel wasp because of its bright metallic green color. It is fairly good-sized, slightly over ¾" long; parts of the second and third pairs of legs are red.

The emerald cockroach wasp was imported from New Caledonia in 1940. This wasp is sometimes seen on walls of buildings searching for prey; usually it is outside hunting prey among vegetation. It is beneficial, since one of its prey is cockroaches. We recently observed a female wasp in action. It had stung and paralyzed a large American cockroach when we came upon the scene. The cockroach was stunned only enough so that it couldn't move on its own. As the wasp towed the cockroach along by its mouthparts (sometimes the wasp will bite off the cockroach's antennae and pull it by the antennal stubs), the cockroach's legs moved to keep the cockroach upright. This is extremely efficient if you are a jewel wasp, because you use your prey's wheels to help move it, instead of carrying the full weight of the prey. The paralyzed cockroach was towed by the little wasp

Figure 114. The emerald cockroach wasp stings and paralyzes her prey, but only just enough so that the cockroach can't use its legs. She then grabs the cockroach by its mouthparts and tows it to her lair, where she lays her egg in it. The larval wasp feeds and develops in the paralyzed cockroach. (Photo by Gordon Nishida)

over gravel, rocks, and up the wall, like a child pulling a wagon. We did not see where the wasp ultimately dragged the cockroach, but once back in her lair, the wasp would have laid a single egg in the cockroach; the resulting larva feeds on the cockroach, eventually making a cocoon inside the cockroach body and emerging as an adult wasp.

If you can't stand this little wasp in your household, shoo it outside. Remember, though, that it can be helpful in reducing your cockroach populations.

Larger Ensign Wasp
Evania appendigaster
Family Evaniidae
(Fig. 115)

Figure 115. The larger ensign wasp when seen from the top seems not to have an abdomen. From the side, the abdomen is very tiny and hatchet-shaped. The female wasp lays her eggs in cockroach eggs and the young hatch and feed on the cockroach eggs. (Photo by Gordon Nishida)

This is a small, black wasp about 1/3" long. The name "ensign wasp" comes from the way the hatchet-shaped abdomen is carried upward like a flag and waved back and forth. This is another household friend that is a parasite of cockroaches; in this case, the wasp lays its eggs in cockroach eggs and the developing wasps feed on the cockroach eggs. The larger ensign wasp is sometimes seen on walls in the house searching for prey.

Paper Wasps
Golden Paper Wasp
Polistes fuscatus aurifer
Common Paper Wasp
Polistes exclamans (Fig. 116)
Macao Paper Wasp
Polistes macaensis
Redbrown Paper Wasp
Polistes olivaceous

Figure 116. The black-and-yellow patterns of the common paper wasp and related wasps warn possible predators that they either taste bad or might sting. (Photo courtesy of Van Waters & Rogers Inc.)

Hawai'i has four species of paper wasps commonly found around houses. They feed on other insect pests, especially caterpillars, and may thus be considered beneficial. They also feed on meat, plant and fruit juices, and scavenge many kinds of food items.

Paper wasps *(Polistes)* look like their more aggressive relatives, the yellowjacket wasps *(Vespula)*; both are dressed in yellow and black or brown warning coloration. *Polistes* wasps are larger (about ¾" long), more slender, and have a spindle-shaped abdomen that tapers at both ends. Yellowjackets are smaller, stouter, and have a broader abdomen. Paper wasps are usually peaceful and shy, but can sting if angered.

Figure 117. Paper wasps chew wood to make paper, out of which they make their nests. They stock the cells in the nest with caterpillars for the young to eat. (Photo courtesy of Van Waters & Rogers Inc.)

Paper wasp nests may be seen hanging from a thin stalk in areas protected from the weather: from eaves of buildings, ceilings, fences, tree branches, shrubs, and other plants. Nests are made from chewed wood and plant fibers mixed with saliva; they look like paper (Fig. 117). The nest has individual, six-sided cells in one layer; the cells are exposed and not covered over with a protective envelope. A single egg is laid by the fertilized queen into each cell; the nest may contain up to 200 cells. When the eggs hatch in a few week's time, the young (larvae) are fed day to day on sweet solutions and caterpillars or other insects. Adults emerge in a few months.

If wasps are not bothering you and are out of the way, leave them alone. Remove the nest later when all wasps have moved out. If the wasps are in an area possibly dangerous for people or pets, control should be directed at the nest, not at flying adults. However, attempts to destroy nests can aggravate adults and cause them to chase and sting people. Stings are painful and can be dangerous to sensitive individuals.

Insecticides for use against wasp nests are commonly available as aerosols that release a strong stream rather than spray; these allow the applicator to stand at a safe distance from the nest. Application should be in the early morning or late afternoon when most adults are on the nest, are less active, and are less likely to attack. If the nest is large or there are many nests, insecticidal application is best left to pest control professionals.

Figure 118. The slender muddauber adult also has a black-and-yellow pattern. (Photo courtesy of Van Waters & Rogers Inc.)

Figure 119. Muddauber nests are the familiar earthern cubicles plastered onto walls. Muddaubers usually stock their nests with spiders to feed their young. (Photo courtesy of Van Waters & Rogers Inc.)

Muddauber
Sceliphron caementarium
(Fig. 118)

The muddauber is a large ($1^1/3$"), black-and-yellow wasp with a very long, thin waist. Muddaubers make nests of mud and stock them with spiders (Fig. 119). They then lay their eggs in the mass of spiders, which become living food for the young. Muddaubers can often be seen around dripping faucets and other sources of water to pick up mud balls that they use to make their nests. They like sticking their nests on walls of houses and similar structures. If angered, they can sting, but they are usually quite shy and retiring.

Yellowjackets
Western Yellowjacket
Vespula pensylvanica (Fig. 120)
Common Yellowjacket
Vespula vulgaris

Figure 120. The western yellowjacket is a very aggressive wasp and is a threat to native insects. The wings of the yellowjacket fold into narrow strips over the back and are expanded when it flies. (Photo courtesy of Van Waters & Rogers Inc.)

A discussion of wasps wouldn't be complete without mentioning the yellowjackets. These aren't household pests, but they may be found in the neighborhood. In Hawai'i, they seem to stay at middle elevations and more often in forested areas.

Yellowjackets are similar to paper wasps, which have the alternating black-and-yellow coloration that warns predators to stay away. See the discussion above under paper wasps for differences between the two groups. Learning to recognize the yellowjackets is important because they are potentially dangerous. They live in large colonies and usually nest in the ground (Fig. 121), though sometimes their nests are aerial. They both bite and sting and, unlike bees, can sting multiple times. If nests are threatened or, sometimes if you are even near a nest, wasps may make a mass attack.

Figure 121. Western yellowjacket nests are often very large, consist of many wasps, and are in the ground. (Photo courtesy of Van Waters & Rogers Inc.)

Currently in Hawai'i, yellowjackets are a serious problem only for the native insects. Yellowjackets catch and feed insects to their young and are very efficient predators. If populations of yellowjackets increase in Hawai'i, they may become bigger problems to humans. In parts of the world, they are extreme nuisances when they search out meat and sugar sources around people, sometimes turning a picnic lunch into a battle zone! Because of their aggressiveness and tendency to attack in large numbers, yellowjackets are probably among the most dangerous of wildlife. People should be on the alert when they are around.

Honey Bees

Honey Bee
Apis mellifera
Family Apidae

Honey bees occasionally become a problem in Hawai'i when a queen and her brood find a swarming or nest site in or around the home. Large swarms of bees can be dangerous, so you should be careful around them. If you are stung by many bees or have a reaction to the sting, seek medical help immediately. If the swarm is not in a location dangerous to humans, leave it alone and it may eventually move away on its own. However, if the bees have established a nest, you may have to seek help. Honey bees are known to nest in house walls, attics, cracks, rocks or concrete blocks, and similar places. Because of the possible danger to you and your family, do not attempt to deal with a bee colony on your own. The Vector Control Branch of the Department of Health can advise you on how to handle your situation. Or, consult a licensed pest control operator. For more information on the health impact of bees, see *What Bit Me?*

SELECTED REFERENCES

Berenbaum, M. 1989. *Ninety-nine gnats, nits, and nibblers.* Urbana: University of Illinois Press. 254 pp.

Gertsch, W. J. 1979. *American spiders*, 2nd ed. New York: Van Nostrand Reinhold Co. 274 pp.

Kingi, T. L., Jr. Undated. *Tropical pests and uninvited guests.* Self published. 20 pp.

Klein, H. D., and A. M. Wenner. 1991. *Tiny game hunting: Environmentally healthy ways to trap and kill the pests in your house and garden.* New York: Bantam Books. 278 pp.

Lifton, B. 1991. *Bug busters.* Garden City Park, New York: Avery Publishing Group. 254 pp.

Mallis, A. 1982. *Handbook of pest control: The behavior, life history, and control of household pests*, 6th ed. Cleveland: Franzak & Foster Co. 1,101 pp.

Metcalf, R. L., and R. A. Metcalf. 1993. *Destructive and useful insects: Their habits and control*, 5th ed. New York: McGraw-Hill. 1,084 pp.

Nishida, G. M. (ed.). 1994. *Hawaiian terrestrial arthropod checklist*, 2nd ed. Bishop Museum Technical Report No. 4. 287 pp.

Nishida, G. M., and J. M. Tenorio. 1993. *What bit me? Identifying Hawaiʻi's stinging and biting insects and their kin.* Honolulu: University of Hawaiʻi Press. 72 pp.

Olkowski, W., S. Daar, and H. Olkowski. 1991. *Common-sense pest control.* Newtown, Connecticut: Twunton Press. 715 pp.

Termite Times. Cooperative Extension Service. College of Tropical Agriculture and Human Resources, University of Hawaiʻi at Mānoa.

Urban Pest Press. College of Tropical Agriculture and Human Resources, Cooperative Extension Service, University of Hawaiʻi at Mānoa. Flyers on control of pest species in Hawaiʻi.

ACKNOWLEDGMENTS

We owe thanks to many individuals who gave us advice on the book and contributed photographs and specimens. We are indebted especially to Laurence Allen, formerly of Van Waters & Rogers Inc. (VWR), West Sacramento, California, for supplying us with many of the color photographs that he produced over his 16-plus years with VWR; we greatly appreciate the generosity of Van Waters & Rogers Inc. in making these slides available as part of their continuing education program.

Dr. Julian Yates III, University of Hawai'i, helped immensely with the termite chapter, and Dr. Neil Reimer, University of Hawai'i, was kind enough to read and comment on the ant chapter. Dr. Frank Howarth and Carla Kishinami of Bishop Museum helped with advice and identifications. Drs. Al Samuelson, Neal Evenhuis, Sabina Swift, David Preston, and Scott Miller, Bishop Museum, were generous with their advice and time.

Others providing photographs were Julian Yates, Neil Reimer, Larry Nakahara, and Ron Heu. Dennis Kunkel, Pacific Biomedical Research Center, provided the scanning electron micrograph of the house dust mite.

Ipo Santos-Bear, Lois Takamori, Al and Shirley Samuelson, and Ben and Keith Leber kindly and enthusiastically shared their pests (and pets) so that they could be photographed. Thanks go to Kay Fullerton, who brought attention to the jewel wasp and its prey. We sincerely apologize if we missed anyone who made this book possible.

Thanks also to the inquiring public, whose many interesting and often challenging questions stimulated the writing of this book.

We are grateful to University of Hawai'i Press for giving us the opportunity to help others better understand the little critters that sometimes share our homes with us.

INDEX

Acari, 120, 166
Acheta domesticus, 116
Acrididae, 114
ACZA. *See* ammoniated copper zinc arsenate
Aedes albopictus, 17, 109
aggravating grasshopper, 115
American cockroach, 86, 87, 170
ammoniated copper zinc arsenate, 160
Amphicerus cornutus, 64
amphipod, 21, 24, 32
Amphipoda, 89
Ampulex compressa, 170
Ananca bicolor, 69
angoumois grain moth, 127
anhydrous calcium carbonate, 72
Anobiidae, 56–58
Anoplolepis longipes, 39, 40
ant, 13, 21, 26, 31, 32, 36–51, 162; Argentine, 38–39, 40, 42, 47; big-headed, 38, 40; biology of, 41; crazy, 38, 40; glaber, 38, 40; kinds of, 37–40; longlegged, 39, 40; Mexican, 39, 40; pharaoh, 37, 38, 40; as predators, 10, 42; queen, 41; soldier, 41; in termite control, 165; tiny yellow house, 37, 40; versus carpenter ant, 41; versus termite, 151; winged, 41; worker, 41. *See also* Hawaiian carpenter ant
antennae, 21
Anthophoridae, 74
Anthrenus flavipes, 66; *scrophulariae*, 67
Arachnida, 120, 136, 166
Araneae, 136
Argentine ant, 38–39, 40, 42, 47
arsenic, 46
arthropod, 19–23; appendages, 21; life cycle, 22–23
Ascalapha odorata, 131
Asian spinybacked spider, 143, 144
Asian tiger mosquito, 109, 110
Attagenus fasciatus, 67; *unicolor*, 67
Australian cockroach, 86

Bacillus thuringiensis israelensis, 111–112
bagworm, grass, 133
bait, 12, 13; against ants, 43, 46, 47; against cockroaches, 85; homemade, 47; for termites, 165

bamboo powderpost beetle, 64
banana spider, 142
barklice, 71
barriers, 45
basaltic termite barrier, 161
bed bug, 33, 52–54
bee, 15, 21, 26; bumble, 74; honey, 74, 174; Sonoran carpenter, 74–76
beetle, 15, 26, 31, 33, 55–70; bamboo powderpost, 64; black carpet, 67; black larder, 68; blister, 69; cabinet, 68; carpet, 67; cigarette, 56–57; click, 69, 70; confused flour, 61; dried fruit, 62; drugstore, 58; false blister, 69; furniture carpet, 66; herbarium, 58; hide, 65, 66, 68; pineapple sap, 62; powderpost, 64; powderpost bostrichid, 64; red flour, 61; sawtoothed grain, 58; wardrobe, 67
big-headed ant, 38, 40
biological control, 16
bird mite, 120–122
bird nest, 66
Bishop Museum, 66
Blaberidae, 88
black carpet beetle, 67
black earwig, 91
black larder beetle, 68
black light, 17
black widow spider, southern, 19, 20, 137
black witch, 131–132
Blattella germanica, 87
Blattellidae, 87
Blattidae, 86
blister beetle, 69
blow fly, 102
Bombus spp., 74
booklice, 26, 33, 71, 123
borate, 12, 63, 161
borer, lesser grain, 59
boric acid, 11, 12, 13, 44, 46, 47, 84, 92, 135
Bostrichidae, 59, 64
Brachycyttarus griseus, 133
brewer's yeast, 99
bristletail, 25, 134
brown dog tick, 166–167
brown recluse spider, 136, 140
brown violin spider, 136, 140–141
brown widow spider, 20, 137, 138

brownbanded cockroach, 87
Bufo marinus, 88
bug, 23, 26, 31, 33, 73; Pacific kissing, 73
bug light, 9–10, 18
bug sucker, 17
bug zapper, 17
bumble bee, 74
burrowing cockroach, 88
butterfly, 21

cabinet beetle, 68
cadelle, 61
Calliphoridae, 102
camphor, 16, 130
Camponotus variegatus, 40, 48
cane spider, 8, 136, 142
carbamate, 14, 101, 169
carbaryl, 99, 100
carpenter ant. *See* Hawaiian carpenter ant
carpenter bee. *See* Sonoran carpenter bee
carpet beetle, 67
carpet moth, 128
Carpophilus hemipterus, 62
casemaking clothes moth, 128
cat flea, 93, 94–95
CCA. *See* chromated copper arsenate
cedar oil, 131
cedarwood, 131
centipede, 8, 16, 17, 21, 24, 33, 77; large, 77–79
cheese skipper, 108
Cheiracanthium diversum, 143
Chelisoches morio, 91
chemical control, of fleas, 97, 98; of termites, 162. *See also* control
Chilopoda, 77
chlordane, 18, 163
chloropicrin, 163
chlorpyrifos, 46
chromated copper arsenate, 160
Chrysanthemum, 100
cigarette beetle, 56–57
Cimex lectularius, 52
Cimicidae, 52
citronella, 15, 16, 100, 113
click beetle, 69
clothes moth, 15, 128–131
cockroach, 13, 23, 25, 31, 33, 80–88, 162; American, 86, 87; Australian, 86; brownbanded, 87; burrowing, 88; German, 83, 87–88; in house dust, 123; Surinam, 88
Coleoptera, 26, 55
collars, pet, 17; flea, 101; ultrasonic, 98

Collembola, 25, 149
Combat, 13, 85
common paper wasp, 172
common yellowjacket, 173
coneheaded grasshopper, 115
confused flour beetle, 61
control, biological, 16; of ants, 45–48; of bed bugs, 54; of carpenter ants, 50–51; chemical, 10–16; of cockroaches, 83–85; commercial, 18; of fabric, carpet, and hide beetles, 65; of filth flies, 104–107; of fleas, 98; miscellaneous, 17–18; of mosquitoes, 111–113; of spiders, 147–148; of storedfood beetles, 56; strategies, 4–7; of termites, 162–165; of ticks, 168–169; of wood and paper pests, 63
Coptotermes formosanus, 155
crab, 89
crane fly, 102
crayfish, 89
crazy ant, 38, 40
cricket, 21, 25, 31, 34, 114, 115–117; flightless field, 116; house, 116; oceanic field, 117
Crustacea(n), 32, 89
Cryptotermes brevis, 154
Ctenocephalides felis, 94
Ctenolepisma longicaudatum, 134
Culex quinquefasciatus, 109, 110
Culicidae, 109
Curculionidae, 60
Cymbopogon nardus, 15

d-Limolene, 97, 99
daddy longlegs, 143
daring jumping spider, 145
deet, 15, 16, 112
Dermaptera, 25, 91
Dermatophagoides spp., 123
Dermestes ater, 67; *maculatus*, 68
Dermestidae, 65, 66
Diacide, 13
diatomaceous earth, 11, 13; against amphipods, 89; against ants, 46, 51; against bed bugs, 54; against booklice, 72; against cockroaches, 84; against earwigs, 92; against fleas, 99; against silverfish, 135; against stored food pests, 56
diazinon, 50
Dictyoptera, 25, 80
diethyltoluamide. *See* deet
Dinoderus minutus, 64
Dipylidium caninum, 95
Diplopoda, 118

dip, 97
Diptera, 26, 102
dog dung fly, 102, 104
dog tapeworm, 95
dragonfly, 16
drain fly, 108
Dri-Die, 12
dried fruit and sap beetles, 62–63
dried fruit beetle, 62
Drione, 12, 100
Drosophila spp., 107
Drosophilidae, 107
drugstore beetle, 58
drywood termite. See termite, drywood
Dursban, 50
dusts, 11–13; against ants, 46; against fleas, 99

earwig, 21, 25, 34, 91–92; black, 91; European, 91; ringlegged, 91; striped, 91
Elateridae, 69
electrogun, 63, 164
emerald cockroach wasp, 170–171
ensign wasp, 8
Ephestia kuehniella, 127
Euborellia annulipes, 91
Euconocephalus nasutus, 115
European earwig, 91
Evania appendigaster, 171
Evaniidae, 171

fabric beetle, 65
false blister beetle, 69
fiberoptic scope, 158
filth fly, 102–107
flea, 21, 26, 34, 93–101; cat, 93, 94–95; human, 93; oriental rat, 93, 95
flea collar, 17, 98, 101
flea trap, 96
flesh fly, 102
flightless field cricket, 116
flour beetle. See grain and flour beetles
fly, 15, 21, 26, 31, 34, 102–108; blow, 102; crane, 102; dog dung, 102, 104; drain, 108; filth, 102–107; flesh, 102; fruit, 107; green bottle, 102–103; house, 102, 103–104; moth, 108; vinegar, 107
fogger, 101
forest tree termite, 154
Forficula auricularia, 91
Formicidae, 36
Formosan subterranean termite. See termite
frass, 50, 63
freezing, 7, 56, 63, 65

fruit fly, 107
fumigant, 14. *See also* repellent
fumigation, 7, 18
fungi, 165
furniture carpet beetle, 66

Gasteracantha mammosa, 143
gecko, 8, 9–10, 16; house, 9
German cockroach, 83, 87–88
glaber ant, 38, 40
Glycyphagus domesticus, 122
golden paper wasp, 172
grain and cereal moths, 126–128
grain and flour beetles, 58–62
granary weevil, 60
grass bagworm, 133
grasshopper, 21, 25, 114, 115; aggravating, 115; coneheaded, 115; long-horned, 114; short-horned, 114; vagrant, 114
green bottle fly, 102, 103
Gryllidae, 115
Gryllodes sigillatus, 116
guppy, 111

harvestman, 143
Hawai'i State Department of Agriculture, 23
Hawaiian carpenter ant, 33, 39, 40, 41, 48–51
heat, in controlling insects, 7, 56, 63, 165
herbarium beetle, 58
Heteropoda venatoria, 142
Heteroptera, 26, 52, 73
hide beetle, 65, 66, 68
honey bee, 74, 174
house cricket, 116
house dust, 123, 124
house dust mite, 123–125
house fly, 102, 103–104
house gecko, 9
house mite, 122
house spider, 147
household casebearer, 133
human flea, 93
huntsman spider, 142
hydramethylnon, 13, 46, 85
Hymenoptera, 26, 36, 170

IGR. *See* insect growth regulator
Incisitermes immigrans, 154
Indianmeal moth, 127–128
insect growth regulator, 99, 100, 101, 165
insect structure, 20–22
insecticides, precautions for use, 11. *See also* control, chemical, and specific chemicals

insects, 25; capturing and killing, 2
integrated pest management, 3
Iridomyrmex glaber, 38; *humilis*, 38
isopod, 24
Isopoda, 90
Isoptera, 25, 150

jewel wasp, 170

katydid, 25, 114, 115
key, to insects, by signs or damage, 27–29; by habitat or part of house, 29–30; by kind of insect, 30–35

Labidura riparia, 91
lacewing, 16
ladybug, 16
large brown spider, 136, 142–143
large centipede, 77–79
larger ensign wasp, 171
Lasioderma serricorne, 56
Latrodectus geometricus, 20, 137; *hesperus*, 137; *mactans*, 19, 20, 137
Lepidoptera, 26, 126
Lepisma saccharina, 135
lesser grain borer, 59
life cycles, arthropods, 22–23
Linepithema humile, 38, 40
Liposcelis divinatorius, 71
lobster, 89
long-horned grasshopper, 114
longlegged ant, 39, 40
lowland tree termite, 154
Loxosceles reclusa, 140; *rufescens*, 140
lufenuron, for flea control, 99
Lyctidae, 64
Lyctus brunneus, 64

Macao paper wasp, 172
maggot, 21, 34, 102
mealworm, yellow, 60
Mediterranean flour moth, 127
methoprene, 100
Metoponorthus pruinosus, 90
Mexican ant, 39, 40
millipede, 21, 24, 34, 77, 118–119; rusty, 118; *Spirobolellus* sp., 118–119
mite, 24, 34, 120–125; house, 122; house dust, 123–125; northern fowl, 120; prevention/control of rat and bird, 122; rat and bird, 120–122; stored-products, 122–123; straw itch, 122; structure, 20, 21; tropical fowl, 120; tropical rat, 120

Monomorium floricola, 37, 40; *pharaonis*, 37, 40
mosquito, 15, 17, 26, 32, 102, 109–113; Asian tiger, 109, 110; southern house, 110; species in Hawai'i, 109–110
mosquito fish, 111
moth, 16, 21, 26, 32, 34, 126–133; angoumois grain, 127; black witch, 131–132; carpet, 128; casemaking clothes, 128; clothes, 15, 128–131; grain and cereal, 126–128; grass bagworm, 133; household casebearer, 133; Indianmeal, 127–128; Mediterranean flour, 127; webbing clothes, 128
moth fly, 108
muddauber, 173
Musca domestica, 103, 104; *sorbens*, 104
Muscidae, 102

naphthalene, 15, 130
Neotermes connexus, 154
nests, carpenter ant, 49; treating, 47–48
Nitidulidae, 62
northern fowl mite, 120

oceanic field cricket, 117
Ochetellus glaber, 38, 40
Oedemeridae, 69
Off! Skintastic, 15
oil films, 112
Oncocephalus pacificus, 73
ootheca, 80
organophosphate, 14, 101, 169
oriental rat flea, 93, 95
Ornithonyssus bacoti, 120; *bursa*, 120; *sylviarum*, 120
Orthoptera, 25, 114
Oryzaephilus surinamensis, 58
Osmose-treated wood, 160

Pacific kissing bug, 73
pale leaf spider, 143
palmetto bug, 86
paper pests. *See* wood and paper pests
paper wasp, 171–173. *See also* wasp
paradichlorobenzene, 7, 15, 16, 130
parasite, 16
Paratrechina bourbonica, 38; *longicornis*, 38, 40
PDB. *See* paradichlorobenzene
pennyroyal, 100
Periplaneta americana, 86; *australasiae*, 86
pesticide, 11. *See also* insecticide
petroleum jelly, 45, 54

pharaoh ant, 37, 38, 40
Pheidole megacephala, 38, 40, 42
Phereoeca allutella, 133
Phidippus audax, 145
Pholcidae, 143, 145
Pholcus phalangioides, 143–144
pillbug, 32, 90
pills, for fleas, 99
pineapple sap beetle, 62
Piophila casei, 108
Piophilidae, 108
piperonyl butoxide, 14
Platorchestia platensis, 89
Plodia interpunctella, 127
Poison Center, 23
Polistes, 172; *exclamans*, 172; *fuscatus aurifer*, 172; *macaensis*, 172; *olivaceous*, 172
Porcellio laevis, 90
powderpost beetle, 64
powderpost bostrichid, 64
powders. *See* dusts
Precor, 100
predators and parasites, 8, 16
Program. *See* lufenuron
protozoan, 150, 152, 155
Pseudomyrmex gracilis mexicanus, 39, 40
Psocoptera, 26, 71
Psychoda spp., 108
Psychodidae, 108
Pulex irritans, 93
Pycnoscelus indicus, 88; *surinamensis*, 88
Pyemotes boylei, 122
pyrenoid. *See* pyrethrins
pyrethrins, 12, 13, 14; against ants, 46; against bed bugs, 54; against carpenter bees, 76; against fleas, 99–100, 101; against mosquitoes, 112, 113; against ticks, 168; against stored products pests, 56; against woodborers, 63
pyrethrum. *See* pyrethrins

rat and bird mite, 120–122
red flour beetle, 61
redbrown paper wasp, 172
repellent, 14–16; against fleas, 100; against mosquitoes, 112; against termites, 163
Resmethrin, 51, 162
Rhipicephalus sanguineus, 166
Rhyzopertha dominica, 59
rice weevil, 60
ringlegged earwig, 91
Roach Prufe, 12
rusty millipede, 118

Salticidae, 145–147
sap beetles. *See* dried fruit and sap beetles
sawtoothed grain beetle, 58
Sceliphron caementarium, 173
Schistocerca nitens, 114
Scolopendra subspinipes, 77
scorpion, 21
screens, 10, 84, 111
Scytodes, 147
Sentricon, 165
short-horned grasshopper, 114
silica aerogels, 11, 12; against amphipods, 89; against ants, 44, 46, 51; against booklice, 72; against cockroaches, 84; against earwigs, 92; against fleas, 100; against stored food pests, 56; against woodborers, 63
Silvanidae, 58
silverfish, 21, 25, 34, 123, 134–135
Simodactylus cinnamomeus, 70
Siphonaptera, 26, 93
Sitophilus granarius, 60; *oryzae*, 60
Sitotroga cerealella, 127
Skin-So-Soft, 16
smoke coil, 112
soap, 45, 48, 54, 96, 97; insecticidal, 48
sodium borate, 161
Sonoran carpenter bee, 74–76
southern black widow spider, 137
southern house mosquito, 110
sowbug, 32, 90, 123
species, 19
Sphecidae, 170
spider, 8, 16, 24, 35, 136–148; Asian spiny-backed, 143, 144; banana, 142; brown recluse, 136, 140, 141; brown violin, 136, 140–141; brown widow, 137; capturing/identifying, 147; cane, 8, 136, 142; control of, 147–148; daring jumping, 145; house, 147; huntsman, 142; large brown, 136, 142–143; pale leaf, 143, 144; pholcid, 145; southern black widow, 19, 20, 137; structure, 20, 21; uloborid, 147; western black widow, 137; widows, 137–140
Spirobolellus sp., 118–119
spray, 14; for fleas, 100–101
springtail, 21, 25, 35, 149
Stegobium paniceum, 58
stored-food pests: beetles, 55–63; mites, 122–123; moths, 126–128
straw itch mite, 122
striped earwig, 91
subterranean termite. *See* termite
sulfuryl fluoride, 18, 162

Supella longipalpa, 87
Surinam cockroach, 88
symbiosis, 151

Tadd Dog, 164
tapeworm, dog, 95
Tapinoma melanocephalum, 37, 40
tarantula, 136, 137, 142
Tegeneria domestica, 147
Teleogryllus oceanicus, 117
Tenebrio molitor, 60
Tenebrionidae, 60, 61
Tenebroides mauritanicus, 61
Term-Out, 51, 162
Termi-Mesh, 162
termite, 8, 18, 21, 25, 32, 34, 150–165; barriers against, 161–162; control of, 162–165; detection of, 157–158, 164; drywood, 18, 150, 152–153, 154, 162; favorable conditions for, 156–157; forest tree, 154; Formosan subterranean, 150, 153, 155–156, 157, 158, 160, 162, 163; lowland tree, 154; prevention of invasions, 159; in shake roofs, 161; swarms, 152; tunnels, 156; versus ant, 42, 151; West Indian, 154, 160; wing, 41
Tettigoniidae, 115
Thelyphassa apicata, 69
Theridiidae, 137, 144
Theridion rufipes, 144–145
Thysanura, 25, 134
tick, 24, 34, 166–169; brown dog, 166–168; structure, 20
Tilapia, 111
Tinea pellionella, 128
Tineola bisselliella, 128
tiny yellow house ant, 37, 40
toad, 16
Toxorhynchites, 109
traps, sticky, 8, 10; flea, 96
Tribolium castaneum, 61; *confusum*, 61
Trichophaga tapetzella, 128
Tricorynus herbarium, 58
Trigoniulus lumbricinus, 118
trilobite, 19
Trogoderma anthrenoides, 68
Trogositidae, 61
tropical fowl mite, 120
tropical rat mite, 120

Uloboridae, 147
ultrasonic collar, 98
ultrasound, 17
ultraviolet light 17, 18
University of Hawai'i Diagnostic Service Center, 23
urban silverfish, 134–135
Urophorus humeralis, 62

vacuum, 17, 18, 66, 84, 98, 123, 140, 168
vagrant grasshopper, 114
Vector Control Branch, Hawai'i State Department of Agriculture, 23, 98
Venezillo parvus, 90
Vespula, 172; *pensylvanica*, 173; *vulgaris*, 173
Vikane, 18, 162, 163
vinegar fly, 107
vitamin B^1, 99

wardrobe beetle, 67
wasp, 15, 16, 21, 26, 32, 102, 170–174; common paper, 171; common yellowjacket, 173; emerald cockroach, 170–171; golden paper, 172; larger ensign, 171; Macao paper, 172; muddauber, 173; redbrown paper, 172; Western yellowjacket, 173; yellowjacket, 172, 173–174
water bug, 86
webbing clothes moth, 128
weevil, granary, 60; rice, 60
West Indian drywood termite, 154, 160
Western black widow spider, 137, 138
Western yellowjacket, 173
white ant, 151
widow spider, 19, 20, 136, 137–140
Wolman treated wood, 160
wood, treatment of, 160–161
wood-boring beetle, 63, 164
wood and paper pests, 63–64
worm, 21

Xenopsylla cheopis, 93
Xylocopa sonorina, 74

yeast, 99
yellow mealworm, 60
yellowjacket, 172, 173–174

Zosis geniculatus, 146, 147

ABOUT THE AUTHORS

JoAnn M. Tenorio is a former entomologist at Bishop Museum and since 1989 has managed the scholarly journals program at the University of Hawai'i Press. She earned her Ph.D. from the University of Hawai'i. Her academic training and interests lie with the medically important insects, mites, and spiders—from which emerged her first book, also coauthored with Gordon Nishida, *What Bit Me? Identifying Hawai'i's Stinging and Biting Insects and Their Kin*.

Gordon M. Nishida has B.A. and M.A. degrees in biological sciences from the University of California at Berkeley and San Jose State University. He is collections manager of Natural Sciences at Bishop Museum, where he coordinates the collections and data management for over 21 million specimens. He is coauthor (with JoAnn Tenorio) of *What Bit Me?*, editor of *Hawaiian Terrestrial Arthropod Checklist*, and author of numerous papers and articles.